H.J Chaney

Our Weights and Measures

A practical treatise on the standard weights and measures in use in the British

Empire. With some account of the metric system

H.J Chaney

Our Weights and Measures

A practical treatise on the standard weights and measures in use in the British Empire. With some account of the metric system

ISBN/EAN: 9783337165093

Printed in Europe, USA, Canada, Australia, Japan

Cover: Foto ©berggeist007 / pixelio.de

More available books at **www.hansebooks.com**

ENTRANCE TO THE PYX CHAPEL.

From the Eastern Cloisters of Westminster Abbey, used as a depository for Standards.

(See page 121.)

OUR WEIGHTS AND MEASURES

A PRACTICAL TREATISE

ON THE

STANDARD WEIGHTS AND MEASURES IN USE

IN THE

BRITISH EMPIRE.

WITH SOME ACCOUNT OF THE METRIC SYSTEM.

BY

H. J. CHANEY.

EYRE AND SPOTTISWOODE,
Government, Legal, and General Publishers.
LONDON: EAST HARDING STREET, FLEET STREET, E.C.
1897.
Price Seven and Sixpence.

TO

SIR COURTENAY E. BOYLE, K.C.B.,

PERMANENT SECRETARY OF THE BOARD OF TRADE.

PREFACE.

No recent account of the weights and measures of the British Empire has been published, and although many official reports and papers on the subject have of late been laid before Parliament, yet, even when accessible to ordinary readers, these reports do not appear to have become so generally known as they might be, particularly so far as they relate to the history of our standards and to local customs.

An attempt has therefore been made in the present treatise to indicate what is now the practice in the use of our various weights and measures either in trade or for the purposes of manufacture, without, however, offering any opinion on questions affecting the construction and administration of the law which governs the practice; and it is hoped that the work may be of some useful help, not only to the manufacturer and trader, but also to the local inspector and others practically interested in the use of weights and measures.

As to the remote origin of our systems of weights and measures much has been written by eminent metrologists, to whose works adequate references may, it is hoped, be found in the present account, and some further historical information has been offered by the present writer.

CONTENTS.

PART I.

1. Origin of the "Imperial System" (p. 1).

Imperial Standards.
Parliamentary Committees, 1758–1821.
Exchequer Standards.
New Imperial Standard Yard and Pound, 1855.
Imperial Standard measures of capacity.
Parliamentary Standards.

2. Ancient Standards (p. 10).

Standards of Henry VII. and Queen Elizabeth.
Standard Wine Gallon, 1707.

3. Local Standards (p. 13).

Local or Inspectors' Standards.

4. Probable Origin of our Systems of Weights and Measures (p. 16).

Eastern origin.
Roman and Anglo-Saxon origin.

5. Units derived from Natural Constants or Physical Standards (p. 21).

Natural Constants or Physical Standards.
Standards from Grains of Wheat, &c.

6 **Standards of** Scotland, Ireland, Channel Islands, **and** Isle of Man (p. 25).

Standards of Scotland.
Standards of Ireland.
   ,,   Channel Islands.
   ,,   Isle of Man.

7. **Standards of the United States** (p. 34).
8. **Standards of India, &c. (p. 36).**
9. **Standards of Canada (p. 42).**
10. **Standards of** Australia, &c. **(p. 44).**

✦ ·●· ✦

PART II.

11. Inspection of Trade Weights and Measures **(p. 46).**

Present practice of the Local Authorities.
Local Practice in Scotland.
   ,,   Ireland.
Denominations of Board **of Trade** Standards.
Board of Trade " Model Regulations."
Ancient Local Practice.
Market Practice.
Courts Leet.
Annoyance Juries.
Merchandise Marks Act.
Inspection by Trade Guilds.
Weights and Measures in docks and harbours.
Standards Commission, 1868–70.
Orders in Council.
Weighing Instruments used in Trade.
Inspectors' Weighing Instruments.

✦ ·●· ✦

LIST OF ILLUSTRATIONS.

FIGS.		Page
Frontispiece.	Entrance to Ancient Depository for Standards.	
1, 2.	Imperial Yard - - - - - -	5, 6
3.	Imperial Pound - - - - -	7
4.	Imperial Gallon - - - - -	8
5.	Bushel - - - - -	9
6.	Winchester Bushel of Henry VII. - -	11
7.	Standard Wine Gallon of Queen Anne -	12
8.	14 lb. Avoirdupois Local Standard (temp. Henry VII.)	15
9.	Ancient Stirling Jug or Scotch Pint - -	28
10.	Scotch Choppin or Half-pint - -	29
11.	Ancient Lanark Standards - - -	29
12.	Inspectors' Local Standards - -	47
13, 14.	Inspectors' Scale-beams - - -	63, 64
15.	Stamp on Inspectors of Gas-meters -	84
16.	Gas-room, Standards Office - - -	85
17.	Comparator for Standards of Length - -	90
18.	Balance-room, Jewel Tower - -	93
19.	Balance of Precision - - -	95
20.	New Metre - - - - -	97
21.	New Kilogram - - - -	98
22, 23, 24.	Representative Forms of Ordinary Metric Weights and Measures - - - -	112
25, 26.	Interior of Pyx Chapel, Westminster -	120
27.	Two ancient Chests - - - -	122
28.	Ancient Keys of Pyx Chapel - -	124
29.	Basement of the Jewel Tower - -	123
30.	Corn "Hopper" - - - -	129
31, 32.	Bushel Measures - - -	130
33.	Miner's Dish - - - -	136
34, 35.	Weights and Measures Offices - -	151, 152
Appendix 2.	Representative Forms of Weighing-Instruments used in Trade - - - -	157

PART I.

1. ORIGIN OF THE "IMPERIAL SYSTEM."

The "Imperial system" of weights and measures now generally in force throughout the British Empire is based on two units, the unit of length or yard measure, and the unit of mass or pound weight, the standards of which are deposited at the Standards Office, 7, Old Palace Yard, Westminster, London.

Imperial Standards.

The legal introduction of the Imperial system into this country dates from 1824, or from the passing of the Act of Parliament, 5 Geo. IV. c. 74, but the description "Imperial" first occurs in the Third Report of the Weights and Measures Commissioners of 1821. The preamble to the Act recites that there were then in use in various places throughout the United Kingdom different weights and measures; that such diversity was the cause of great confusion and manifest fraud, and that for the remedy of these evils for the future, certain standards should be established throughout the United Kingdom.

From the time of Henry III. the Statute Book abounds indeed with Acts providing for uniformity, so forcible has custom been with the people; and even at the present time, notwithstanding the best intentions of the Legislature, we have not quite obtained uniformity throughout the United Kingdom.

Probably it was not only the want during the last century of a uniform system of weights and measures in the internal trade of the country which brought about the legislation of 1824, but also the growing demand for higher accuracy in scientific research, a demand which helped in France to develop the introduction of the metric system.

Parlia-
mentary
Com-
mittees,
1758 to
1821.

The legislation of 1824 was indirectly the outcome of important reports made by Committees of the House of Commons, particularly the report of a Committee of 1758 of which Lord Carysfort was chairman. The latter Committee recommended in their reports of 1758 and 1759 that a "troy pound" (5,760 grains), and a yard measure (36 inches), made according to their directions, should be adopted as new standards, but their recommendation received no legal recognition until 1824. Previously to 1824, and from 1588, the legal standards in use had been those made during the reign of Queen Elizabeth, and until 1588 the standards in use had been those of Henry VII.

Much of the legislation of 1824 was also based on the recommendations of Committees of the House of Commons, and on the reports of Commissioners of Weights and Measures, appointed by the Treasury (1814–21), particularly on the recommendations of a committee of 1816, who examined Dr. W. H. Wollaston, the Secretary of the Royal Society, and Dr. Playfair, F.R.S.; and on the recommendations of commissioners appointed to ascertain the standards of the country, who reported in 1819, 1820, and 1821.

Exchequer
Standards.

In 1825 the Treasury Commissioners called in the assistance of Dr. T. Young, F.R.S., Captain H. Kater, F.R.S., and other eminent scientific authorities, under whose directions brass copies of the new Imperial Standards, or Secondary Standards, were made and deposited at the office of the Receipt of Exchequer, and they became known as "Exchequer Standards." Copies of the Exchequer Standards were also sent during the same year to the three Metropolitan Cities, London, Dublin, and Edinburgh. The Exchequer Standards of 1825 are now included with the Board of Trade Standards. Some copies of the Exchequer Standards are still in the custody of the Town Clerk at the Guildhall, London; the Edinburgh copies are in the Museum of Weights and Measures at the City Chambers, Edinburgh; and the Dublin copies are deposited at the City Hall, Dublin.

The Standards legalised in 1824 were injured or destroyed at the burning of the Houses of Parliament in 1834, they having remained in the custody of the Clerk of the House of Commons since 1758–60 in accordance with a resolution of the House dated 2nd June 1758. The legal life of the first Imperial Standards was, therefore, short as compared with the old Winchester Standards, which lasted from 1588 to 1824. The yard of 1760 was constructed by Bird, the optician, and was based on a brass yard constructed by Graham for the Royal Society in 1742, which was derived from the yard measure of Queen Elizabeth. The Troy Pound of 1758 was originally verified by Mr. Harris, the King's Assay Master, by comparison with the Exchequer Troy Weights of 1601. After the burning of the Houses of Parliament certain of the Standards of 1758–60 were found in the ruins, particularly the yard measure of 1760. These ancient standards have now, under the care of Sir R. Palgrave, K.C.B., been placed in the lobby of the residence of the Clerk of the House of Commons leading from the official corridor at the back of the Speaker's chair.

Destruction of Standards in 1834.

In 1838, at the instance of the Chancellor of the Exchequer (Right Hon. T. Spring Rice, subsequently Lord Monteagle), a Commission was appointed to consider the steps to be taken for the restoration of the standards. The Commission made a report in 1841, and in consequence of their report a Committee was appointed in 1843, who reported in 1854, and new Imperial Standards were made under their directions, which standards were duly legalized by an Act of 1855 (18 & 19 Vict. c. 72). The new Act substituted the *Avoirdupois* pound of 7,000 grains for the Troy pound of 5,760 grains. The yard and pound of 1855 were verified by comparison with standard weights and length - scales, which had been compared with authenticated copies of the original standards of 1758-60 by the Rev. R. Sheepshanks, F.R.S., Mr. F. Baily, F.R.S., and Professor W. Hallows Miller, F.R.S. The scientific methods by which the legal standards of 1855 were verified have been described in two classical papers, published in the Philosophical Transactions of the Royal Society of London in 1856-7. There appears to be no doubt that the present Imperial Standards have been accurately derived from those of Queen Elizabeth, and that these latter were derived from those of Henry VII.

Copies of the new Imperial Standards were deposited in 1854 in the Houses of Parliament, at the Royal Observatory, in the Royal Mint, and with the Royal Society. Such copies are legally known as " Parliamentary Copies," and in event of the original standards at the Standards Office being lost or injured, new standards can, under the Weights and Measures Act, 1878, be created by reference to or by adoption of such Parliamentary Copies.

The Standards Act of 1855, was repealed by the Weights and Measures Act, 1878, now in force, but so much of it as particularly described the existing Imperial Standards was re-enacted in the latter Act.

With the view of giving effect to the provisions of the Standards Act of 1866, which transferred the custody of the standards from the Comptroller - General of the Exchequer to the Board of Trade, a Royal Commission was appointed by Warrants, dated 9th May 1867, and 4th May 1868, to inquire into the condition of the Exchequer Standards, and to direct and superintend the steps necessary to be taken until all these official standards were proved to be in perfect condition (see page 58).

The Imperial Standards are particularly described in the First Schedule to the Act of 1878. The yard is a solid square bar (Figures 1 and 2), made of bronze or gun metal, on which is marked the length, at the temperature of 62° Fahrenheit, of the Imperial Standard yard of 36 inches.

Description of the present Imperial Standards.

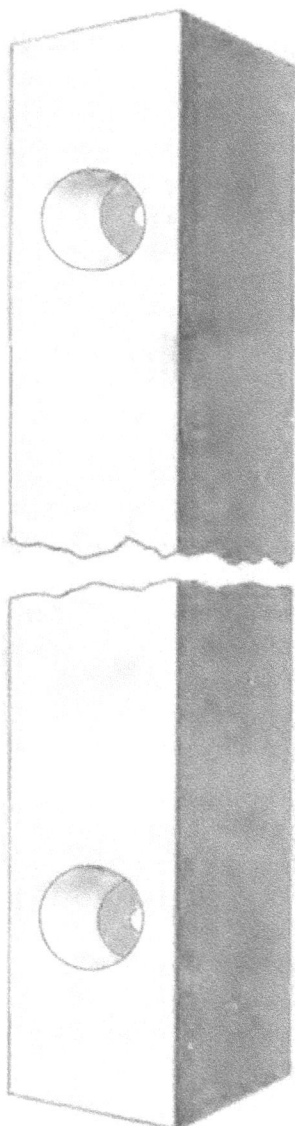

FIG. 1.
Present Imperial Standard Yard (No. 1).

B

The following is a plan and section of the Imperial Standard Yard bar:—

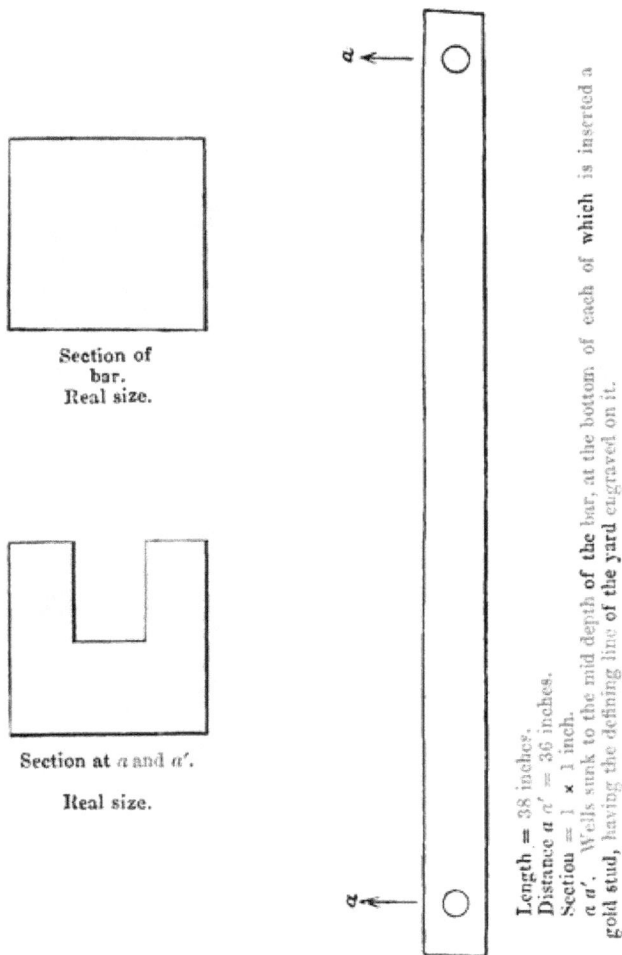

Section of bar.
Real size.

Section at *a* and *a'*.

Real size.

Length = 38 inches.
Distance *a a'* = 36 inches.
Section = 1 × 1 inch.
a a'. Wells sunk to the mid depth of the bar, at the bottom of each of which is inserted a gold stud, having the defining line of the yard engraved on it.

FIG. 2.

The bronze is composed of copper 16, tin 2½, and zinc 1, ounces avoirdupois.

The **Imperial Avoirdupois Pound** is made of platinum, and is in the form of a cylinder nearly 1·35 inches in height, and 1·15 inches in diameter, with a groove or channel round it for insertion of the points of the ivory fork, by which it may lifted, and it is marked " P.S. 1844. 1 lb." Its form is shown in Fig. 3.

Avoirdupois is our principal weight, as the Act of 1878 requires all articles sold by weight **to be** sold by avoirdupois, excepting **that** gold and silver, and all precious metals or stones **may be** sold by the ounce, troy ; and drugs when **sold by retail may** be sold **by** apothecaries' weight.

Upper Surface of Pound.

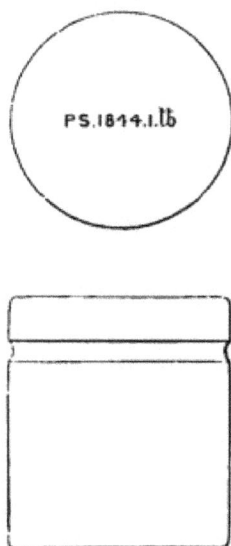

Fig. 3.—PRESENT IMPERIAL **STANDARD POUND.**

The present unit of capacity for liquids as well as for dry goods is the Imperial gallon measure, introduced in 1824 in place of the old Winchester gallon.

Imperial Standard measures of capacity.

It is determined by a brass gallon measure (Fig. 4), which, like all standard measures below the gallon, has its diameter equal to its depth, and which contains ten Imperial pounds weight of distilled water weighed in air against brass weights, with the water and the air at the temperature of 62° Fahrenheit, and with the barometer at 30 inches. Captain Kater gives an account of the verification of this and other standards of capacity in the Philosophical Transactions of the Royal Society of London, for 1826. The Imperial bushel, or 8 gallons, was verified in 1825, and is made of gun metal (Fig. 5), but, like the other dry measures, the half-bushel and peck, its diameter is double its depth, which proportion was originally selected as affording the truest measure for corn.

Fig. 4.—PRESENT IMPERIAL STANDARD GALLON.

Brass gallon marked "Imperial Standard Gallon Anno Domini MDCCCXXIV. Anno V.G.iv Regis."

Fig. 5.—PRESENT STANDARD BUSHEL.

As the Imperial gallon contains ten pounds weight of water, it can be taken that, throughout the British Empire, " A pint of pure water weighs a pound and a quarter."

By the Act of 1878, the Parliamentary copies of the Imperial Standards are required to be compared with each other once in every ten years; but those immured at the Houses of Parliament are only compared with the Imperial Standards once in every twenty years, and they were so last compared in 1892 as described in a report laid before Parliament (Commons Paper, No. 171 of 1892). The place where the immured standards are deposited is indicated by a brass plate on the right-hand side of the first landing-place on the steps leading up to the Committee Rooms of the House of Commons. At the proceedings with reference to the examination of the Parliamentary Standards on 2nd April 1892, at the Palace of Westminster, there were present the Right Hon. A. W. Peel, the Speaker; Sir Michael Hicks Beach, M.P., President of the Board of Trade; Sir H. G. Calcraft, Secretary of the Board of Trade ; the Right Hon. D. Plunket, First Commissioner of Her Majesty's Works;

Parliamentary Standards.

Colonel W. Carrington representing the Lord Great
Chamberlain, and Mr. H. J. Chaney, Superintendent of
Weights and Measures.

————•◦•——— —

2. ANCIENT STANDARDS.

Standards
of Henry
VII. and
Queen
Elizabeth.

The ancient standard of linear measure appears to
have been re-established in 1324 by Statute of Edward II.
(17 Edward II.), which ordained that three barley-corns,
round and dry, make an inch, 12 inches a foot, three
feet a yard, or *ulnam* (First Report Commissioners of
Weights and Measures, 1819).

By other statutes and decrees, particularly the
Statute 31, Edward 1, 1303 (Tractatus de Ponderibus
et Mensuris ; see also Statute of 25 Edward 1 Magna
Carta, c. 25), provision had been made for maintaining
uniformity of measure. The Statute of 1303 appears
to be the first in which the unit of linear measure was
particularly referred to.

In the reign of Henry VII. several Statutes were
passed relating to uniformity of measure (1491,
7 Henry VII. ; 1495, 11 Henry VII. ; 1496–7,
12 Henry VII.).

The standard yard of 36 inches (1496) still exists,
and is probably of the same length as the old Saxon yard.

This ancient standard is a bronze rod, the length of the
yard being the distance between the ends of the rod, and
it was in constant use for the verification of other yards
until the reign of Queen Elizabeth, when a new standard
was made (1588), which measure is also at the Standards
Office (*see* Descriptive List of ancient standards. Paper
presented to Parliament by Command. C.—6541. 1892.)

The standard bushel of Henry VII. (1495) is shown in Fig. 6.

With reference to the Winchester bushel (Henry VII.), it may be interesting to note that although the ancient Winchester bushel is now illegal in trade, its use for fixing corn rents is not obsolete. For instance, in Lincolnshire, three arbitrators were appointed in 1885 by the justices to ascertain the average price of a Winchester bushel of good marketable wheat for the past 21 years, to form the basis for re-ascertaining the corn rents for the next 21 years.

Three separate gallon measures had been in use as follows from ancient times (two of them probably from 1225) until 1824, when, by the passing of the Act, 5 Geo. 4. c. 74, the present imperial gallon, then containing 277·274 cubic inches, was introduced, and the use of the three ancient gallon measures was made illegal.

	Cubic Inches.
A Winchester corn gallon -	$= 272\frac{1}{4}$
A wine gallon - -	$= 231$
An ale gallon - - -	$= 282$

Fig. 6.—WINCHESTER BUSHEL OF HENRY VII. (1495).

This bushel agreed in capacity with Queen Elizabeth's bushel (2150·4 cubic inches).

Illustrations of other ancient obsolete standards have been given in the 7th Annual Report of the Warden of the Standards for 1873.

Standard
Wine
Gallon,
. 1707.

Fig 7.--STANDARD WINE GALLON OF QUEEN ANNE, 1707. (⅓ size.)

A new standard measure, made of bronze, was deposited in the Exchequer in 1707, of which Fig. 7 is an illustration (*see* " United States," p. 35).

3. LOCAL STANDARDS.

The standards hitherto described belong to the State and (*inter alia*) are used in verifying other subsidiary standards known as Local Standards or Inspectors' Standards, by which are controlled the trade weights and measures—as those used in shops, &c. Such Local Standards are required by the Act, of 1878, to be verified by the Board of Trade; the weights once every five years and the measures once every ten years. Such comparison is made on the application of any local authority requiring it, either at the Standards Office, Westminster, at the Royal Irish Constabulary Depôt, Phœnix Park, Dublin, the General Assembly Hall, High Street, Edinburgh, or elsewhere.

(margin: Local or Inspectors Standards.*)*

The form and material of all local standards are set out in detail in an official paper (A 3.—1885) issued to Local Authorities. Generally the standards are made of brass, bronze, or gun-metal; the measures of capacity, from a bushel to a quarter-gill, are of cylindrical form, as shown on the right hand side of Figure (p. 47); the avoirdupois weights (56 lbs. to ½ dr.) are either bell-shaped, or of spherical shape, as shown in the middle of the illustration (Fig. 12).

A list of denominations of trade weights of the Imperial system is given in Appendix I.; but local inspectors are generally supplied with the following standards of trade weights and measures:—

Avoirdupois weights—56, 28, 14, 7, 4, 2, 1, lbs.; 8, 4, 2, 1, ounces; 8, 4, 2, 1, ½, drams.

Troy weights, decimal series, 500 ounces down—10, 5, 4, 3, 2, 1, 0·5, 0·4, 0·3, 0·2, 0·1, to 0·001 ounce troy.

Decimal grain weights—4,000, 2,000, 1,000, 500, 300, 200, 100, 50, 30, 20, 10, 5 to 0·1 grain.

Apothecaries' weights—10, 8, 6, 4, 2, 1, ounces; 4 drachms to $\frac{1}{2}$ grain.

Measures of capacity—Bushel, $\frac{1}{2}$ bushel, peck, gallon, $\frac{1}{2}$ gallon, quart, pint, $\frac{1}{2}$ pint, gill, $\frac{1}{2}$ gill, $\frac{1}{4}$ gill.

Apothecaries' measures—From 10 fluid ounces to 10 minims (glass).

Measures of length.—A yard, subdivided into feet and inches, the inch being divided into 12 duodecimal, 10 decimal, and 16 binary equal parts.

In large manufacturing centres the local authorities provide other denominations of Inspectors' Standards required in their district.

When standards are verified for the use of a local authority, an indenture or certificate of verification is issued bearing a seal. The old Indenture Exchequer seal was engraved " Sigillium officii recepte scaccarii regio in Anglia " (Official seal of the Receipt of the King's Exchequer). The present Indenture bears the seal of the Board of Trade; and all local standards have stamped on them the Board of Trade stamp—Crown, Sovereign's initials, date of verification, and the portcullis.

Since the reign of Queen Elizabeth the local standards have always been provided out of local rates or taxes, and not by the State. In ancient times, however, local standards were provided by the State; thus Henry V. (Exchequer Record. Roll Issue. Michaelmas. 9 Henry V) (see Devon's Issues of the Exchequer. London, 1837, p. 371), ordered 100 various weights to be made for the King's benefit and use of his people. The Act 11 Henry VII. c. 4 (1495) required that one of every weight and measure, according to the Exchequer standard made of brass, should be delivered to 43 cities and towns named in the Act. The following is an illustration of a weight supplied in 1495; it bears the words " Henricus Septimus " in high relief with the

Tudor Rose on one side and the Westminster portcullis on the other :—

Fig. 8.—14 lb. AVOIRDUPOIS WEIGHT (temp. HEN. VII. (⅓ size.)

In the reign of Queen Elizabeth, also, by a Royal Proclamation, Roll of 17th June 1588, standards were ordered to be delivered to 58 cities and towns named therein. In ancient times the old standards were generally melted down to make new standards, and hence, so far as can be ascertained, none of the local standards in use before those of Henry VII. are now in existence. A collection of old local standards at Winchester is interesting, as it includes standard troy weights dated 1588, being the year in which Elizabeth granted a charter to Winchester, and a Winchester bushel sent to the Corporation from the Exchequer in 1487. The old " Winchester Bushel " was so called because the standard bushel was ordered by King Edgar to be kept there. The collection at Winchester also includes other standards of 1487, 1601, 1700, and particularly a 56 lb. weight, supposed to be of the time of Edward III., which was found in the old muniment room over the West Gate. There is also an interesting collection of old local standards at the museum at Norwich.

An Act of 1670 (Charles II.) enacted that a standard bushel of correct measure should be chained in the market place of all market towns in order that persons might test their measures.

◆◆◆

4. PROBABLE ORIGIN OF OUR ANCIENT SYSTEMS OF WEIGHTS AND MEASURES.

Eastern origin.

Our units of weights and measures, like those of other European countries, appear to have come from the East, through Greece and Rome, and their origin is almost pre-historic.

In Holy Writ, we have the earliest references to the use of weights and measures; as for instance, Genesis XXIII., 16; 1 Kings VII., 15; 2 Chron. III., 15; Isaiah XL., 12.

In estimating the measures of the past, if we believe that the great Pyramid of Jeezeh, which still exists, and which was built before the art of writing was communicated to mankind, was really designed for the purpose of perpetuating the standards of measure by its astronomical proportions and bearings, as Mr. John Taylor ("Battle of the Standards," 1864), and Professor Piazzi Smyth, F.R.S. ("Life and Work at the Great Pyramid," 1867), would almost convince us, and that our measures have been derived from those of the Great Pyramid; then, our measures have an origin more remote than those of any other country.

Dr. Arbuthnot ("Tables of Ancient Coins," 1754), says Josephus "tells you that Cain was the first monied man, " that he taught his band luxury and rapine, and broke

" the public tranquility by introducing the use of weights
" and measures."

Queipo ("Systèmes Métriques et Monétaires des anciens
peuples." Par Don V. Vasquez Queipo, Paris, 1859),
appears to show that there were in the East three grand
ancient systems—the Assyrian, Egyptian, and Phœnician—
and that from these came the old European systems.
Although the classical study of the origin of our
metrological systems begins in the East, we cannot say
with certainty from which of the most ancient countries
our principal units have been each derived. Some writers
have strung together a chain of coincidences, and inductive
reasoning has been strained to show that our units are of
Mosaic origin. Our units, however, hardly appear to have
such an origin; certainly our systems of weights and
measures were not selected by those acquainted with
arithmetic and astronomy: probably our present systems
have grown with the exigencies of our expanding trade
and commerce.

Our measures of length appear to be partly Roman and
partly Anglo-Saxon. The "old London mile" of 5,000
feet of 1,000 paces was, doubtless, derived from the
Roman; and the yard from the Saxon. Professor Greaves
("Origin and Antiquity of our English Weights and
Measures," 1745, p. 68), states that the Anglo-Saxons
were a free people from the beginning, and scarce so much
as felt the power of the Romans, and that, although since
their coming into Britain they were overpowered by the
Danes and Normans, " it cannot seem improbable that they
" should have preserved their most ancient measures."

Bishop Fleetwood says ("Chronicon Preciosum," 1707,
p. 34), that it was a good law of King Edgar that there
should be the same weight and the same measure throughout
the kingdom. Probably the Saxons brought their own
weights and measures along with them from Germany.

Roman and Anglo-Saxon origin.

Mr. Donisthorpe (" Measures, Past, Present, and Future,
London, 1895 "), remarks that the inch, foot, and mile, were
brought over by the Romans and forced upon the British
in spite of the fact that they bore no commensurable
relation to the then existing measures of length.

The Romans used the libra, which, like our Troy pound,
was divided into 12 unciœ or ounces, and the later Greeks
appear to have had their litra, which they divided after
the same manner. Some say also that the Romans determined their unit of capacity, the congius, by its weight
of water—ten pounds, like the Imperial gallon.

Between our measures of capacity and our weights
there has always been some relation, and there is a close
resemblance between some of our measures of capacity
and the Roman measures. The ancient English pint, for
instance, is so close to the Roman sextarius (1·01 pint)
that one can hardly doubt the derivation of the pint from
the sextarius. In like manner the old Scots pint (see " A
Bushel of Corn," by A. Stephen Wilson, Edinburgh, 1883,
pp. 3 and 4) was so close to the Roman semi-congius " that
" it is impossible to doubt the genetic identity of elements
" in the two measures." In the Domesday Book mention
appears to be made of the sextar.

Our weights appear to have come, some from Rome,
some from Arabia, and some are Anglo-Saxon.
Dr. Arbuthnot ("Tables," 1754) shows that the Roman
ounce was similar in weight to the English avoirdupois
ounce; and also that the " Pondo argenti " amongst the
Romans was a numerical expression of sums of money, and
was different from the common Libra (lb.). Dr. Langwith,
however ("Observations on Dr. Arbuthnot's Dissertation,
London, 1754 "), has pointed out that the Norman French
name of the weight (avoirdupois), and the fact of its being
divided into 16 ounces and not 12 ounces show that the
avoirdupois pound was not of Roman origin.

As to troy weight, some have derived the word from Troyes in France, an ancient commercial centre, and some have derived the word from Trone weight, as **according to** Fleta (c. 1340), "trona" meant a beam to weigh with.

There appears to be no trustworthy record of **the** origin of the Troy system. The Troy pound is said **to** have been derived from the Roman weight of 5759·2 grains the 125th part of the **large** Alexandrian talent (Third Report, Standards Commission, 1870, p. 3). The earliest statute in which Troy weight is named is the 2 Henry V. s. 2. c. 4—"pur la libre de Troy orre."

Lord Swinton, in his work "A Proposal for Uniformity of Weight and Measure in Scotland," 1789, states, page 133, that it is not improbable that "Troy" or "Troye" weight **derived its** name from trois, the Norman for three, **the** original weights being, he states, the penny, shilling, **and** pound, or the penny, ounce, and pound. Wilson, however (pp. 14–16), is of opinion that troi weight, **or trois weight,** was derived from droit pois (25 Edward III. Stat. **5. c. 9**), or right weight.

According to the Statute "Tractatus de Ponderibus **et** Mensuris," there were two pounds in use, a legal pound **of 12 ounces and** a merchants' **pound of** 15 ounces. In **a Statute,** 9 Hen. V. **c.** 11, there is notice of even a third pound, called **the** Pound of the Tower (London), or Moneyers' Pound, which appears to have been 360 troy grains less than the legal pound of 12 ounces, and the **use** of which appears to **have** been abolished in 1527. **The** Tower Pound **is** supposed by Snelling ("Silver **Coins,"** 1762, p. 4) to have been the same as that used **by** the Saxons, and to have contained **12** ounces **of** 20 pennyweights each, the first **of which sub-divisions** was used by the Romans.

Professor W. H. Miller, **F.R.S.** (Phil. Trans., Roy. Soc., 1856-7, **p.** 753), remarks that "the earliest legal standard

" of English weight, of which any very authentic account
" is preserved, is the weight called the Pound of the
" Tower of London. According to Folkes it was the old
" pound of the Saxon Moneyers before the Conquest."

Mr. H. W. Chisholm states (Seventh Annual Report,
1873, p. 31), that the exact period of the introduction of
avoirdupois weight into this country is not known, and
that there is evidence tending to show that standards of
avoirdupois weight existed in the time of Edward III.

Ruding ("Annals of the Coinage," 1817) presumes,
however, that the Anglo-Saxons regulated their coins by
a pound kept at the Tower, that being then the chief seat
of the Mint, and he assumes they brought their pound with
them from Germany.

Of the avoirdupois ounce, Bishop Cumberland ("Essay
on the Jewish Weights and Measures." London, 1686,
p. 74) remarks that it is the same as the Roman, the
subdivision of the pound into 12 ounces following the
practice of the Egyptians, Greeks, and Romans. It does
not seem at all clear when the avoirdupois pound was first
divided into 16 ounces, and at one time it appears to have
been divided into 15, and 20, parts or oræ.

Some have concluded that the ounce avoirdupois was
descended from the Jewish shekel, and Bishop Cumberland
(l.c. pp. 11 and 105) seems to say that the Roman ounce
and the avoirdupois ounce had their " true original " in
the double of the Eastern sanctuary shekel. Our system
of weights has been by some traced back to the Babylonian
" mina" by means of ancient weights now in existence, and
which date back to 2000 B.C. One of the most interesting
of these Maneh or Mina weights has been referred to by
Professor Flinders Petrie, F.R.S., and is now deposited at
the British Museum ; it is a green stone weight, of conical
form, found in Egypt, and it weighed 15097·6 grains.
It bears an interesting cuneiform inscription, which has

been translated as follows :—" One maneh standard weight,
" the property of Merodach-sar-ilami, a duplicate of the
" weight which Nebuchadnezzar, King of Babylon, the
" son of Nabopolazzar, King of Babylon, made in exact
" accordance with the deified weight of Dungi, a former
" king." The system of ancient Assyrian and Babylonian
weights, from which the Greek system of weights has been
considered to be derived (as mina or maneh), has been
described by Layard (" Discoveries in the Ruins of Nineveh
and Babylon," 1853), and the derived weight of one mina
appears sometimes to have approximated to the weight of
an avoirdupois pound.

Apothecaries' weights, like apothecaries' measures,
appear also to have had a Grecian and an Arabic origin.
Carat weight is an instance of an ancient Arab weight ;
being formerly known as karact and caract weight.

------- ◆•◆•◆ -------

5. UNITS DERIVED FROM NATURAL CONSTANTS OR PHYSICAL STANDARDS.

Our standard of measure or of weight is not now Natural
required by law, if lost or injured, to be restored by constants
or physical
reference to any natural constant or physical standard, standards.
but by reference to material or arbitrary standards constructed by hand. The "yard" measure is no longer
required, as it formerly was (1824) to be restored if
lost, by reference to the length (39·1393 inches) of the
" Pendulum vibrating seconds of mean time in the latitude
" of London in a vacuum at the level of the sea." The
"pound" weight also is no longer to be restored if lost
by reference (1824) to the weight of a " cubic inch of
" distilled water at the temperature of 62° Fahrenheit."

o

The Standards Act of 1855 declared that the researches of scientific men had thrown doubts on the accuracy of such methods of reference to constants in nature. The actual loss of the means for restoring the Imperial Standards appears, in fact, the most unlikely of all possible contingencies.

In like manner the theoretical metre or unit of metric measure is no longer recognised as the one ten-millionth part of the elliptic quadrant of the meridian passing through Paris, but simply as the length marked on a metal bar which is deposited with the International Committee of Weights and Measures at Paris—the "Metre-International." So also the theoretical kilogram or unit of metric weight is no longer derived from the weight of distilled water contained in a cubic decimetre, but is the mass represented by a weight made of iridio-platinum metal, and which is also deposited at Paris—"Kilogramme-International."

Besides the pendulum, some have proposed as natural constants of measure, the space through which a body "falling freely from quiescence" will descend in a given time at a given place; or the length given by an open pipe or tube when yielding a determinate musical sound, as in China, where the measure known as the Ch'ih appears to have been originally obtained from the keynote of the ancient musical scale.

Professor De Morgan was of opinion that to assure to posterity for, say, 500 years hence, the knowledge as to what our yard measure really was, the standard of it must be something which could not be altered by man either from design or accident ("Elements of Arithmetic," 1840).

Sir John Herschel ("The Yard, the Pendulum, and the Metre," 1863) has proposed that the earth's polar axis should be referred to; a proposal, perhaps, that has not received so much public consideration as it might have done. The earth's diameter also has been proposed as a standard of measure.

The most important inquiry, however, of late years with reference to a physical constant standard measure appears to be that made by Professor A. A. Michelson (Valeur du mètre en longueurs d'ondes lumineuses. Paris, 1894). Professor Michelson has found that the metre (39·37 inches) contains 1,553,163·5 wave lengths of the red ray of the spectrum of cadmium, measured in air at 15° centigrade and under an atmospheric pressure of 760 millimetres, and that the actual length of the metre bar at Paris can be determined by this method to within a micron.

Although, however, in many countries attempts have been made to base standards of measure on some physical constant, or standard, yet hitherto no such attempts have received legal recognition.

The proportions of the human body have, in former ages, been largely referred to, e.g., the *foot* measure ; the digit, or finger's breadth ; an inch, or thumb breadth ; the nail, or from the tip to the middle joint of the longest finger ; the palm, as measured across the middle joints of the four fingers ; the hand, clasped with the thumb uppermost ; the span, thumb and little finger extended to the utmost ; the cubit, a length from the elbow to the extended finger ; the step ; the pace, or two steps. Some have derived the *yard* from a Saxon word signifying the girth of the king's body ; and the ell from *ulna*, as used in Magna Charta, " Composito ulnarum et perticarum." The foot and nail are still legal measures ; and the " hand " now used in measuring horses was originally declared, by a Statute of 27 Henry 8, c. 6., to be equal to 4 inches.

By law, units of weight and measure have been based on certain substances, such as seeds, &c. One of our earliest statutes (51 Henry 3, 1266, " Assiza Panis et

Standards from Grains of Wheat, &c.

Cervisiæ ") provided that the English penny, called the
" sterling," should weigh " thirty-two grains of wheat, well
" dried, and gathered out of the middle of the ear;" " that
" twenty pence do make an ounce, and twelve ounces a
" pound." The expression " sterling," as followed in the
above Act of 1266, has been considered by some to have
been a possible origin of the word *sterling*, as applied to
money at the present time; but according to Jeake
(" Arithmetick," 1701, page 76) English money is often called
sterling money because there was anciently stamped upon
it a little bird, called a starling; although other writers
think the coin was minted at a place called Esterling.
Snelling (Gold Coins, 1763, p. 1), quoting from Maitland's
History of London, 1739, also states that in 1257 the King
caused to be coined in London a penny of fine gold,
" weighing two sterlings," which is supposed to be the
first gold coin in England.

The **Statute of 17** Edward II. (1324) had also provided
that " three barley corns, round and dry, made an inch;"
and the earlier Statute 33 Edward I. (1305) had provided
that an inch contained "three barley corns laid end to end."
It would appear that even now the wild hillsmen of
Annam weigh their gold dust by grains of maize and rice.
Dr. Edward Bernard (" De Mensuris et Ponderibus," Oxford,
1685), states that many of the ancients served themselves
with ordinary grains of corn for the measures, both of
length and capacity.

We have, too, a legal weight of a *stone* (14 lbs.) and
formerly we had in Scotland the Lanark stone; and in
Ireland the Belfast stone, for flax, of 16¾ lbs.; all probably
derived from some mass of stone kept under local
governing authorities of a remote period.

The **cow**, or **ox**, has been **used by** primitive people in
measuring **value, and Professor** Ridgeway (" Origin of
Metallic Currency and Weight Standards," Cambridge, 1892)

has pointed out that, by taking as our primitive unit the cow or the ox, we may be able to give a more simple account of the genesis of the ancient weight standards than that which has hitherto been the received one, the value of the cow forming the first gold unit in the following proportions :—

Homeric Ox Unit = 130 to 135 grains of gold.
Roman Ox Unit = 135 grains of gold.
Sicilian Ox Unit = 135 grains of gold.
Ancient German Ox Unit = 120 grains of gold.

In Ireland, for instance, in the reign of Richard II., there appears to have been but little money, and the Irish then "merchandised chiefly with cattle" (Ruding).

---

6. STANDARDS OF SCOTLAND, IRELAND, CHANNEL ISLANDS, AND ISLE OF MAN.

The same Imperial standards of weights and measures are adopted throughout Scotland as are adopted in England. The Act of Union passed in 1706 provided that the English standards only should be used, yet it was not until the year 1835 that the use of the old Scotch standards (as well as of all local and customary weights and measures) was made an offence punishable by fine. Some of the old Scotch measures, as the "Scotch acre" (1·26118 Imperial acre) as well as the "boll" (containing 8857·289 cubic inches for wheat, and 12,921·222 cubic inches for barley, Edinburgh measure), the "forpit," or fourth part of a peck are, indeed, still in use in the measurement of grain and meal. Before the legal intro-duction of the Imperial system into Scotland (1824), the standards then in use not only differed among themselves, but rarely represented what were then the legal standards of the country.

The old Scotch standard included the Scottish Troyes pound, being one-sixteenth part of the Lanark stone, or 7,609 Imperial grains, known generally, also, as Dutch weight; and also the Tron one pound weight, generally used for butter, cheese, meat, hay, and other commodities. The Tron pound varied considerably in weight, but in Edinburgh it contained 9622·67 Imperial avoirdupois grains (Buchanan's "Weights and Measures." Edinburgh, 1843, p. 21).

In 1835 it was ordered that "Fiar Prices" (or Feus), in Scotland, by which the average price in each county of different kinds of grain was annually fixed for the purpose of determining rents and prices in sales, feus, or leases, should be converted into Imperial weights and measures. The Fiar prices of all grain in every county were to be struck by the Imperial Quarter (5 & 6 Will. 4, cap. 63, s. 16). Such Fiar prices are still annually or periodically struck in Scotland (see Paterson's "Historical Account of the Fiars").

Generally, the Scotch pint, or the stoup of Stirling, was taken as the liquid standard of measure. This measure was defined in the Scotch Act of 1618 as being "3 pounds "7 ounces of French Troyes weight of clear running water "of Leith." Looking at the present condition of the waters of Leith, it is difficult to believe that even a pint of clear water should ever have been found there.

The standard of capacity for wheat appears to have been the Edinburgh firlot, containing 2214·32 Imperial cubic inches; whilst the Scotch standard for barley, oats, and malt, appears generally to have been the firlot kept at Linlithgow, equal to 3230·30 cubic inches.

Old Troyes, or Lanark, **weight** :—

16 drops	-	= 1 ounce.
16 ounces	-	= 1 pound.
16 pounds	-	= 1 stone.

The Scotch tron weight used for Scotch produce particularly, had the same scale as the Troyes, but the 64 pounds were equal to 88 avoirdupois pounds.

DRY MEASURE.

4 lippies	-	= 1 peck.
4 pecks	-	= 1 firlot.
4 firlots	-	= 1 boll.
16 bolls	-	= 1 chalder.

The Scotch barley firlot is taken as equal to 815,516·442 Imperial avoirdupois grains, **weight** of distilled water at 62° Fahrenheit. The wheat firlot was taken as 559,023·3675 grains, or nearly **an** Imperial bushel.

LIQUID MEASURE.

The Stirling jug **or pint** contained 26,306·982 grains' weight of distilled **water at 62°** Fahrenheit (Buchanan, p. 201). The English **wine gallon, also formerly used in** Scotland, **contained, in 1826,** 0·8331109 Imperial gallon.

In actual use retail **traders had, therefore, the firlot, peck,** gallon, and forpit as dry measures; and as liquid measures the gallon, $\frac{1}{2}$ gallon, quart, pint, choppin, mutchkin, $\frac{1}{2}$ mutchkin, gill, $\frac{1}{2}$ gill, and $\frac{1}{4}$ gill. The Scotch "choppin" appears to have been derived from the French wine measure the "chopine," identical with the German **and Swiss** "schoppen."

MEASURES OF LENGTH.

The ell was the unit of length, and was equal to 37·0598 English inches, the Scotch mile = 1,976·522 English yards ; 48 Scotch acres being nearly equal to 61 Imperial acres.

The magistrates of Edinburgh have formed a collection at the Municipal Buildings of the ancient Scotch standards. The collection includes the Stirling jug or pint, Fig. 9, and the Scotch choppin of **1555** (Fig. 10); the corn half-forpit (1783); the mutchkin standard of 1767; the gill (1737); a set of Lanark stone troy weights (1618), Fig. 11; and the Scotch ell standard of 1663, which formerly was fastened by a chain to the top of the wall of the Council Chamber at Edinburgh.

In the Council Chamber at Linlithgow there are also preserved a number of ancient Scotch standards; but the Linlithgow standard firlot was destroyed in the fire which consumed the Town House in 1847 (Waldie's " History of Linlithgow," 1879).

Fig. 9.—ANCIENT STIRLING JUG OR SCOTCH PINT.

Fig. 10.—SCOTCH CHOPPIN OR HALF PINT, 1555.

LANARK ONE POUND.

Fig. 11.—LANARK STONE TROY WEIGHT.

The stone does not bear the date 1618, and is probably of earlier date. Other weights are marked 1618.

Standard "Cran."

By the Act 52 & 53 Vict. c. 23, to amend the Herring Fishery (Scotland) Acts it is provided that the "cran" or "quarter-cran" may be used in the measurement of fresh herrings, but sale by weight or in number or by bulk is not prevented, and such measures, made of wood or other material as the Fishery Board of Scotland shall direct are to be branded or otherwise marked according to the regulations of the Board, who have at Edinburgh a standard of this measure. The cran equals $37\frac{1}{2}$ Imperial gallons.

Standards of Ireland.

The same system of weights and measures is followed in Ireland as in other parts of the United Kingdom, the Imperial system having been introduced into Ireland in 1824. In 1495, by the Irish Act 10 Henry 7. c. 22, all English Statutes concerning the public weal were thenceforth to be deemed good and effectual in Ireland (4th Report, Standards Commission, 1870, p. 253), and it would appear that the English standards were indeed legalised so early as 1495. One of the earliest legislative provisions on record affecting the measures of Ireland is contained in the Irish Act 32 Hen. 6. c. 3., passed in 1450, but it is applicable to liquors only. It enacts that no man shall sell wine, ale, nor any other liquor within any city or town franchised, unless it be with the King's measures, sealed, viz. :—

The Gallon,	Pint, or
Pottle ($\frac{1}{2}$ gall.)	Half-pint ;
Quart,	

The gallon referred to is not defined by the Irish Act, and was probably the English wine gallon. Ruding states ("Annals of the Coinage," Vol. I., page 327) that Henry the Second was invited into Ireland "in his seventeenth year" by Dermod Mac Murrogh, King of Leinster, when he settled colonies in different parts of the island, and, by degrees, introduced therein the laws and customs of England. It is probable, therefore, that English weights and measures were accepted in Ireland so early as the twelfth century. In 1695 the Irish Parliament passed an Act (7 Will. 3. c. 24) ; and this Act was followed by another in 1705 ; providing for the deposit in the Irish Exchequer of a set of just and true avoirdupois weights from 56 lbs. to 1 oz., to remain as the standard weights of Ireland. The latter Act also provided for the appointment of "weigh-masters" to execute the duties connected with the local copies of the standards and the commercial weights. The Act also provided that all salt and meal should be sold by weight, and this was extended in 1733 by another Act of the Irish Parliament under which all sorts of corn were to be sold by weight. Special legislation with reference to weights and measures in Ireland took place in 1860 (23 & 24 Vict. c. 119) ; in 1862 (25 & 26 Vict. c. 76), and in 1867 (30 & 31 Vict. c. 94, Dublin Metropolis). The Acts of 1860 and 1867 were repealed by the Consolidated Weights and Measures Act, 1878. The Act of 1862 was partly also repealed by the Act of 1878, but part of that Act appears to be still now in force (Part III., and also so much of Part IV. of the Act of 1862 as relates to Part III.), so far as relates to the prevention of fraud in branding or stamping at a market or fair to denote the weight, measure, or quality of any article sold in the market or fair.

The general practice in Ireland as to standards is similar to that in other parts of the United Kingdom, but sections 76 and 77 of the Act of 1878 specially provide as to the use of Imperial weight in the sale of corn, grain, pulses, potatoes, hay, straw, flax, roots, carcases of beef or mutton, wool, or dead pigs. The mode of weighing also is regulated ; every article sold by weight, if weighed, is to be weighed " in full net standing beam" and no deduction or allowance is to be claimed or made for "tret or beamage" or for the weight of any sack, vessel, or other covering or on any other account.

In Ireland the practice as to the local inspection and appointment of officers is referred to at page 49 (Inspection of Weights, &c.).

The Reverend Isaac Warren states (" Table and Formula Book," Longmans, 1889,) that Irish Long Measure agrees with English, as follows :—

> Irish perch   =   7 yards.
> English perch =   5½ yards.

Hence 11 Irish miles = 14 English miles, and 121 Irish acres = 196 English acres.

Channel Islands. In the Channel Islands there are in use three systems of weights and measures—the Imperial, Metric, and the old Norman systems. The weights and measures of the Channel Islands are regulated by the local Legislature, and are not specially referred to in the Imperial Weights and Measures Acts. Copies of the Imperial Standards were verified at the Exchequer at Westminster in 1844 for the use of the Island of Jersey, and were delivered into the custody of the Viscount of Jersey.

Weights.—Jersey pound avoirdupois of 16 ounces = 7,561 grains Imperial, the same weight as the " livre de la Vicomte de l'Eau," or the ancient Standard of Rouen (Marc de Rouen).

Jersey 52 lbs. avoirdupois, and a nest of Jersey Standards from 52 lbs. to ⅛ oz.

Measures of capacity.—The estendard du chasteau or cabot = 4 gallons, 1 quart, 3 gills, Imperial, or 10 Jersey pots, as stated in an Act of the Jersey Court on 19th January 1625.

The Jersey quart = $\frac{1}{20}$th estendard du chasteau.

There appear to be also in use Jersey measures of a pot, pint, half-pint, noggin, and half-noggin, as well as the following measures :—

		Pots.	Pints.	Noggins.
Large cabot for barley	-	12	3	$1\frac{1}{3}$
Measure for oysters	-	56	0	0
„ potatoes and coals	-	9	$2\frac{1}{2}$	0
„ apples	-	0	2	0
The sixtonnier	- -	2	0	$2\frac{2}{9}$

Measures of length.—**10 ft.** rod for inspection of roads. The aune or ell **of 4 feet.**

The **Jersey pound of 16 ounces,** and the cabot, were legalised by Acts of **the** Royal Court of **Jersey** on 11th December, 1593, 7th March, **1617, and 19th** January, 1625, which were confirmed by the **Sovereign in Council** in 1717.

A specimen of the Jersey pound avoirdupois **or the** ancient Marc de Rouen; together with the parts **of the** marc cup shape, **and** standard weights **of** the **following** denominations **are deposited at** the Standards **Office :—**

1 livre,	4 gros,
4 once,	2 „
2 „	1 „
1 „	1 demigros.

Copies **of** the Imperial standards were originally verified at the Exchequer for the Isle of Man in 1840, and were delivered to the custody of the Clerk of the Rolls, but, under an Act of Tynwald, 1880, the standards adopted in the Isle of Man, and the modes of inspecting trade weights, are now made similar to those prescribed by the Imperial Weights and Measures Act, 1878. On the application of the Registrar of the Isle of Man the standards of the Isle are to be compared with **those** of the Board of Trade once in every 20 years, and they were last so compared in 1883.

In the sale of corn, &c. in the Isle, the following equivalents
were followed :—a bushel of wheat or rye to consist of
64 lbs. ; of barley 56 lbs. ; of oats 42 lbs. ; of peas or beans
60 lbs. ; of potatoes 56 lbs. ; a boll of wheat or rye to
consist of four bushels, or 256 lbs., and a bushel divided
into four pecks or kischens.

The regulation of weights and measures in the Isle of
Man is in the hands of an officer called a Regulator of
Weights and Measures for the whole island, and four other
inspecting officers, each of whom has sub-standards in use,
verified by the Regulator of Weights and Measures under
a local Act of Tynwald (or local Parliament of the Island).

7. STANDARDS OF THE UNITED STATES.

United
States.

The history of the standards of the United States has
been so intimately connected with the history of the
standards of Great Britain, that some particular reference
to the United States may not be out of place here.

The standards of length in the United States are the
yard and metre. It appears from a bulletin (No. 26, dated
5th April 1893), issued by the United States Coast and
Geodetic Survey, Washington, with the approval of the
Secretary of the United States Treasury, that in future the
" International Prototype Metre and Kilogramme" (deposited
at Paris), will be regarded in the United States as the
fundamental standards of " Length and Mass," and that
the yard and pound in America will be derived from these
metric standards (Board of Trade Annual Report, 1894).

The yard originally was the length between the twenty-
seventh and sixty-third inches, marked on a Troughton
82 inch brass scale, deposited at Washington, and this was

supplemented by a copy (No. 11) of the new Imperial standard of 1855, which was presented to the United States Government by the British Government in that year.

The standard of weight is stated to be the brass troy pound copied from the Imperial troy pound by Captain Kater in 1827, for the United States Mint at Philadelphia, and preserved in that establishment. The commercial standard pound avoirdupois was derived from the troy pound, but in 1855 a copy (No. 5) of the present Imperial pound avoirdupois was also presented to the United States Government, and is deposited at Washington.

The unit of capacity for cereals in the United States is the old Winchester bushel of 2150·42 cubic inches; or 0·96944 Imperial bushel. The original standard gallon by which the United States measure appears to have been determined was the Queen Anne's standard wine gallon of 1707. It contained 231 cubic inches, and was stated to hold 8·3389 pounds avoirdupois of distilled water at the temperature of 39°·83 Fahrenheit. A weighing recently made of the actual contents of this gallon measure made its capacity equivalent to 8·3292 lb. avoirdupois of distilled water at the temperature of 62° Fahrenheit and under a pressure of 30 inches of the barometer. (See Fig. 7, p. 12.)

Copies of the above standards and of parts and multiples of those standards, arranged in a decimal and binary series, were deposited in the several States and at the Custom Houses. There appears to be no public general statute applying to the whole of the States, each State and territory making its own laws; and although each State has its own laws as to weights and measures, yet there would appear to be a general uniformity of practice throughout the States. For instance, as to the sale of corn, we gather from the report in 1891 of the Select Committee of the House of Commons on the Sale of Corn (table prepared by the Clerk of the United States Treasury Department, corrected up to 1891, by Mr. W. Almon Wolff) that 31 out of the 34 States adopt a uniform weight of 60 lbs. to the bushel (Winchester) of wheat.

The American Metrological Society of New York, which was organised in 1873, has published much information with reference to the different weights, &c. in use throughout the States.

Besides the " Imperial " foot of 12 inches, there appears to be in occasional use in the United States an old Dutch foot, derived from the " Amsterdam foot " (11·147 English inches). There were in use in Holland, before the introduction of the metric system, two standards of the "foot "— the " Amsterdam " and the " Rhineland " foot (divided into 12 inches and equal to 12·356 English inches).

By an Act of Congress, approved in July. 1866, the use of the weights and measures of the metric system is made permissible ; and all contracts are declared not to be invalid because the weights and measures expressed or referred to therein are weights and measures of that system.   Metric standards to be furnished to each State ; by a Joint Congressional Resolution of the same date the Secretary of the Treasury was "authorised and directed to " furnish each State with one set of the standard weights " and measures of the metric system." Tables of equivalents are also recognised in the Act.

---

8. STANDARDS OF INDIA, &c.

India.

In India native weights and measures are mostly used in trade as hereafter shown ; but the legal standards are based on those of the Imperial system.

In 1889 the "Measures of Length Act" received the assent of the Governor-General, by which the British yard was adopted as the legal standard of length for India. The metric system was introduced by the Indian Act, No. XI., of 1870, and the Act XXXI. of 1871, but does not appear to have been established for use in trade. The Act of 1889 extends to the whole of British India. A copy of the Imperial yard is required to be kept at Calcutta, and any measure having stamped thereon or affixed thereto a certificate, purporting to be made under the authority of the Governor-General in Council or of a local government, shall be deemed to be correct until its inaccuracy is proved. It is also provided that there shall be kept by the Commissioner of Police in the town of Calcutta, under section 55 of the Calcutta Police Act, 1866; by the Commissioner of Police in the city of Madras, under section 32 of the Madras City Police Act, 1888; and by the district magistrate under section 20 of Regulation XII. of 1827 of the Bombay Code, certified measures of the standard yard, standard foot, and standard inch.

The Government of India has been provided (1889) with an accurate copy of the Imperial standard yard, marked as follows :—

"Accurate copy of Imperial Standard Yard. 1889."

" Calcutta."

"Standard Yard at 85 degrees Fahrenheit."

The commercial length of the yard at 85° Fahrenheit, as well of the lengths of the standard foot, standard inch, and the decimal, duodecimal, and binary parts of the inch, respectively, are also marked.

D

The Assay Master of Calcutta, and the Assay Master at Bombay, have also been provided with verified standard weights of the following denominations :—

$$
\left.
\begin{array}{cccc}
1000, \ 600, & 300, & 200, & 100, \\
60, & 30, & 20, & 10, \\
6, & 3, & 2, & 1, \\
0{\cdot}6, & 0{\cdot}3, & 0{\cdot}2, & 0{\cdot}1, \\
0{\cdot}06, & 0{\cdot}03, & 0{\cdot}02, & 0{\cdot}01, \\
\multicolumn{4}{c}{0{\cdot}006, \ 0{\cdot}003, \ 0{\cdot}002, \text{ and } 0{\cdot}001.}
\end{array}
\right\} \text{Tolas.}
$$

A Tola is equivalent to 180 grains avoirdupois.

Also one each of 60 lb., 30 lb., 20 lb., 10 lb., 6 lb., 3 lb., 2 lb., 1 lb., 6 oz., 3 oz., 2 oz., 1 oz., 10 dwt., 5 dwt., 3 dwt., 2 dwt., and 1 dwt., all troy ; and one each of 6, 3, 2, 1, 0·6, 0·3, 0·2, 0·1, 0·06, 0·03, 0·02, and 0·01 grains.

The Assay Office of Bombay is also provided with verified avoirdupois standards.

As the Imperial pound has its mass in vacuo at 0° C. correction has to be made for the weight of air displaced by the pound, and from the information furnished by Major **Gerald Martin,** of the Bombay Mint, and Lieut-Col. J. Scully, of the Calcutta Mint, the **following data have** been adopted for calculating the **weight of air displaced by** standard weights in India :—

Mean temperature - -	$= 85°$ Fahrenheit.
Mean pressure at 32° Fahr. -	$= \mathbf{29{\cdot}8}$ inches.
Vapour tension - -	$= 0{\cdot}75$ inch.
Latitude - -	$= 22° \ 35' \ 6''{\cdot}5.$
Height above mean sea level	$= \mathbf{22{\cdot}6}$ feet.
g Calcutta - - -	$= g_{45°} \ 0{\cdot}9982515.$
Weight of a litre of dry air at the temperature of 85° Fahr. B. = 29·80 inches	$= 1{\cdot}14820$ grams.
Weight of air displaced by W, the commercial avoirdupois pound of **7,000** grains ($\triangle$ W = 8·143) -	$= 0{\cdot}99016$ grain.

The denominations and equivalents of the native weights and measures are so numerous and varied that a list of them could not be given here. Much information, however, with reference to them will be found in a Report by General Sir R. Strachey, made to the Government of India in the year 1867 ; also in Noback's Munz, Mass-und Gewichtsbuch (Leipzig, 1879); in Dr. W. A. Browne's Merchants' Handbook (London 1879); and in Jackson's Metrology (London 1882).

Generally it may be said that in Northern India the usual unit of weight is the tola, which is the weight of the current rupee coin. The seer is a given number of tolas, varying from 70 to 100. The mun or maund is usually 40 seers. A weight of 5 seers called Pusseree is generally recognised, and the seer is subdivided into 16 chittacks. The rupee of the British Government weighs 180 grains ; the British Government seer equals $2\frac{2}{35}$ lbs. avoirdupois, and the maund equals $82\frac{2}{7}$ lbs. avoirdupois. Local seers and maunds vary on either side of 2 lbs. and 80 lbs. avoirdupois In Southern India the maund usually contains 40 common cutcha seers, the seer being equivalent to 24 current rupees. At Madras the commercial maund is taken at 25 lbs. avoirdupois, divided into 8 viss. In Jugerat a seer of 40 local rupees' weight, a maund of 40 such seers, and a candy of 20 maunds, are the common weights. These maunds vary from 37 to 44 lbs. and the seers are about 1 lb. In Malwa a seer of 80 local rupees' weight and a maund of 20 such seers are common. At Bombay the old seer was about 10 or 12 ozs. avoirdupois, equal to 30 pice. The Bombay maund being reckoned at 40 such seers or nearly 28 lbs., which maund is also used on the Malabar coast, but the seer is the Madras one of 24 rupees' weight. At Bombay and in the Deccan the subdivision of the seer is into 72 parts called tank. The

Deccan seer is commonly 80 of the local rupees, or about 2 lbs., and the maund varies greatly.

Measures of capacity are little used in Northern India ; in Bengal and Southern India they are more commonly used. The usual linear measures are the hath and the guz, divided into 16 girahs or 24 tussoos. The hath varies from 14 to 20 inches, the guz from 28 to 40 inches. Thirty-three inches is the length assumed for the guz in fixing the official land measures of the North-West Provinces. The coss may be 4,000 guz, about $2\frac{1}{4}$ miles, and sometimes half that distance. Measures of area are commonly based on the hath or guz.

Straits Settlements. In the Straits Settlements the native trade weights are used as well as the Imperial weights:—

1 pikul.

$\frac{1}{2}$ pikul.

14, 7, 4, 2, 1, $\frac{1}{2}$, katies.

4, 2, 1, $\frac{1}{2}$, tahils.

8, 4, 2, 1, chees.

8, 4, 2, 1, hoons.

10 hoons = 1 chee.

10 chee = 1 tahil ($1\frac{1}{3}$ oz. avoirdupois).

16 tahil = 1 kati ($1\frac{1}{3}$ lb.)

100 kati = 1 pikul ($133\frac{1}{3}$ lb.)

Burmah. In Burmah, so far as can be at present ascertained, the following measures and weights are still in use, but there appear to be no recognised standards of these native denominations.

Distances are described as "a call," or about 200 yards ; "the sound of a gunshot," or half a mile ; a "stone's throw," or from 50 to 60 yards ; " breakfast distance," that is as far as a man could walk between sunrise and breakfast time, *i.e.*, eight to nine o'clock ; as a "mat," *i.e.*, a quarter of a "taing," or half a mile ; a "moo," or the eighth of a taing, "nga-moo" (literally five great moo), or half a "taing."

The weights in use are :—

- 1 kyeng-rwe = 1 **seed of** the *abrus precatorius.*
- 2   „   = 1 rwe-gyee.
- **4 rwe-gyee** = 1 pai-gyee.
- **2 pai-gyee** = 1 moo.
- **2 moo** = 1 mat.
- 4 mat = 1 kyat.
- 100 kyat = 1 piet-tha (**3·652 lb. avoirdupois**).

The measures **of** capacity depend upon the " teng " or basket, which varies in almost every district. Each **teng** is thus sub-divided :—

- 1 **teng**   = 4 **tsiet.**
- 1 **tsiet**   = 4 pyee.
- 1 **pyee**   = 2 khwet.
- 1 **khwet**   = 2 tsa-lay.
- 1 tsa-lay   = 2 **la-may.**
- 1 la-may   = 2 **la-myek.**
- 1 la-myek = 2 **la-moo.**

An endeavour **has been** made **to introduce a standard** " basket," containing **2,218·19 cubic inches, but it has not** been very successful for want of legislative **authority.**

The measures of length **are** :—

- 1 lek-thit = 1 finger **breadth.**
- 8 lek-thit = 1 maik (or span).
- 3 maik   = **1** toung (cubit, 19½ inches).
- 4 toung   = **1** lan (fathom).
- 7 toung   = **1** ta.
- 1,000 ta   = 1 taing (nearly two English **miles**).
- 6,400 ta   = 1 **yoo-za-na.**

The authorities administering the Governments in Borneo, Borneo, Ceylon, and Hong Kong, **are also** provided with &c. copies of the Imperial standards, although native weights and measures are also used.

Cyprus.

In Cyprus, by a law enacted on 20th May 1890 (No. XI., 1890), the standards are specified as follows :—

Capacity.	Weight.	Length.
2 pints = 1 quart.	400 drams = 1 oke.	12 inches = 1 foot.
2⅖ quarts = 1 Cyprus litre.	1⅖ okes = 1 Cyprus litre.	2 feet = 1 pic.
4 quarts = 1 gallon.	44 okes = 1 Cantar.	3 ft. or 1½ pic = 1 yard.
8 gallons = 1 kilé.	180 okes = 1 Aleppo Cantar.	33 pics = 1 chain.
Liquid Measure, 9 quarts } = 1 kouza.	800 okes = 1 ton.	2,640 pics = 1 mile.
16 Kouzas = 1 load.		

——— ✦•✦•✦ ———

9. STANDARDS OF CANADA.

Canada.

In Canada (Quebec, Ontario, New Brunswick, Nova Scotia, Prince Edward Island, Manitoba, British Columbia, N.W. Territory, &c.) the same units of weight and measure are adopted as in Great Britain and Ireland. The "Dominion Standards" of the yard and of the pound avoirdupois are deposited at Ottawa (and Newfoundland) under the Canadian Act 36 Vict. c. 47. and 48, as described in the First Report, dated 20th January 1874, of the Commissioner of Inland Revenue on the Inspection of Weights and Measures and Gas. These standards are similar in form to the Imperial Standards. Besides these standards the Standards Office at Ottawa is provided with

"Departmental Standards," including avoirdupois standards of a decimal series as follows :—

lbs.	lb.
50	1
30	0·5
20	0·3
10	0·2
5	0·1
3	down to 0·001 pound.
2	

and it is this series (and not the binary series, 56, 28, 14, 7 lbs., &c.) which is mainly adopted for use in trade in Canada. As the outcome of the Report of a Select Committee of the Canadian Senate, appointed on 21st March 1870, the present system of inspection and verification of weights and measures in Canada has been organised. Regulations with reference thereto being issued by the Standards Branch of the Inland Revenue.

A Canadian Act, dated 19th April 1884 (47 Vict. c. 36), to amend the Weights and Measures Act of 1879, contains new provisions with reference to the examination of the weights of canned or packed goods, and to the sale of weights and measures. Further detailed regulations have been also recently issued by the Inland Revenue Department respecting the kind of balances, &c. which traders may use in Canada, such regulations being in place of those given in a Consolidated Order in Council of 1882, and in the instructions to inspectors formerly issued by the Inland Revenue Department.

The metric system was rendered permissive by 34 Vict. c. 24. (Canada), 1871, which was re-enacted by the Consolidating Act of 1873, and the Act gave equivalent values of Imperial and metric weights and measures, made the system legal in contracts and dealings, and empowered the Governor-General to provide standards and to make regulations for the verification of metric weights and measures.

In British Columbia the Ordinances regulating weights and measures are the same as those for Canada, chapter 104 of the Revised Statutes for Canada, 1886, the short title of the Dominion Act being Weights and Measures Act, 42 Vict. c. 16.

---

10. STANDARDS OF AUSTRALIA, &c.

Australia and Africa.
The Imperial system of weights and measures is in force in Australia and South Africa, the several Governments having verified copies of the standards deposited with the Board of Trade, and a local inspection of trade weights and measures is provided for by statutes which govern weights and measures in the Colonies.

The following is a list of the Colonial Acts and Ordinances :

CAPE OF GOOD HOPE.—Act No. 11 of 1858, "An Act " for regulating Weights and Measures in the " Colony of the Cape of Good Hope."

NATAL. — Act No. 11 of 1852, "Ordinance for " establishing Imperial Weights and Measures." Section 11 of Act No. 39 of 1884 refers to weights and measures, and confers certain powers on local boards respecting weights and measures.

NEW SOUTH WALES.—No. 34 of 16 Victoria, 1852 "An Act to amend the Law relating to Weights, " and Measures."

NEW ZEALAND.—No. 30 of 1868, "The Weights and " Measures Act, 1868." No. 27 of 1877 ; section 9 of this Financial Arrangements Act refers to weights and measures.

SOUTH AUSTRALIA.—No. **349** of 48 & 49 **Vict.** (1885), " An Act respecting Weights and Measures."

TASMANIA.—4 William 4 No. 3, " An **Act** for " establishing Standard Weights and Measures " and **for** preventing the use of such as are false " or deficient."

54 Vict. No. **29** contains Regulations **for the** measurement of Fruit Cases.

VICTORIA.—Consolidating Act of 1890, " An Act to " Consolidate the Law relating to Weights and " Measures."

WESTERN AUSTRALIA.—3 William IV., No. **2** of 1833, " An Act for regulation of Weights and Measures."

18 Vict. No. 18, " An Act to Revise and Amend certain parts of Ordinance No. 2 of 1833."

Section 55 of 40 Vict. No. **13** refers incidentally **to** weights and measures.

In the West Indies including Bermudas, **Jamaica,** West Trinidad, Windward **Islands** (Barbados, St. **Vincent,** Indies. Grenada, Tobago, St. **Lucia, &c.**), Bahamas Islands, British Honduras, British Guiana, **Falkland Islands, the Imperial** system is also legally recognised. **In Mauritius,** Seychelles, Ascension, St. Helena, Gambia Settlements, Sierra **Leone,** and the **Gold** Coast the Imperial **system** is also legally recognised, but native weights and measures are also used, particularly in the three first-named places.

So far we have endeavoured to show the origin **of the** standards on which our present systems of weights and measures have been based, and the relation of the standards of the United Kingdom to those adopted abroad in the great Colonies and dependencies. We may now follow the application of these standards to the practical measurements of trade and to the development **of** the present system of inspection of weights and measures.

PART II.

11. INSPECTION AND VERIFICATION OF WEIGHTS AND MEASURES IN USE IN TRADE.

Present practice of the local authorities. The important duty of verifiying and inspecting weights and measures in use in trade is carried out by the Local Authorities, as the County and Borough Councils, who appoint the Inspectors of Weights and Measures. The two principal Acts, which at present control the local inspection, are the Weights and Measures Acts of 1878 and 1889; besides which Acts there are the Weights and Measures Acts of 1892 and 1893, which are hereafter referred to, as well as other laws which relate to some particular use of weights and measures, as—

The Sale of Bread Acts, 1836 and 1838.
The Coinage Acts, 1870 and 1891.
The Factories and Workshops Act, 1878.
The Coal Mines Regulation Act, 1887.
The Weighing of Cattle Acts, 1887 and 1891.

As all public general statutes can be obtained from the Queen's Printer, or may be readily referred to in public libraries, &c., the author has given references only to the titles or dates of Acts quoted. It is evidently undesirable to give extracts, however copious, when the original laws themselves are accessible, and can be readily referred to in event of important legal proceedings arising.

Fig. 12.—INSPECTOR'S LOCAL STANDARD WEIGHTS AND MEASURES.

(See pages 13 and 47.)

The intervention of the State appears generally to have been limited to providing and maintaining the primary standards, and to the verification of copies of these standards for the use of the local authorities. The carrying out of the laws for securing a common agreement between the local standards and the weights and measures used in trade appears always to have been, except in the more ancient times, the duty of the local administration.

Some reference has been already made to the denominations of standards at present used by the inspectors (see p. 13), and it may be interesting to give here a general view of a set of local standard weights and measures of capacity as provided for the use of an inspector. See Figure 12.

The old statute books abound with laws touching the inspection of weights and measures, but it was not until 1795 that any systematic inspection was established, when "Examiners" of weights and measures were required to be appointed by the justices throughout the United Kingdom (Act 35 Geo. 3. c. 102, subsequently extended by the Act 37 Geo. 3. c. 143); and further powers for a similar purpose were given to corporate towns in England and Wales by an Act of 1815 (55 Geo. 3. c. 43). These examiners appear to have been the first local officers who gave up their whole time to the duties of their office. In 1835, by the passing of the Act 5 & 6 Wm. 4. c. 63 (which repealed an Act of the previous year) the system of local inspection was further developed, and separate officers, called "Inspectors" were required to be also appointed by the justices.

In 1859 the Act 22 & 23 Vict. c. 56 provided for the first time that the town council of any borough having a separate court of quarter sessions should also have power to appoint inspectors, or examiners, a power which was, by the Act 24 & 25 Vict. c. 75, extended to all boroughs having a separate commission of the peace. The substance

of these Acts was included in the consolidating Weights
and Measures Act of 1878.

At the present time the local authorities in England
and Wales, who carry out the law with regard to the
detection and prevention of fraud and the stamping of
weights and measures, are the county and borough councils
created by the Local Government Act, 1888 (England and
Wales). Thus the local officers whose duty it is to inspect
are now required to be appointed by local authorities
responsible to the ratepayers. Where a borough has a
population of less than 10,000, its local authority cannot
by the Act of 1888 appoint an inspector. In 1893 an Act
was passed (56 & 57 Vict. c. 19.) to relieve any borough,
which is not a county borough and which has not a
separate court of quarter sessions, from contributing
towards the expenses incurred in the execution of the
Weights and Measures Acts by the council of the county
in which the borough is situated. In England and Wales
there are 434 local inspectors.

Local
practice in
Scotland.

In Scotland the county authorities who have to ad-
minister the Weights and Measures Acts are the county
councils. Until the passing of the Local Government
(Scotland) Act, 1889, the justices in general or quarter
sessions assembled were the local authorities, and in some
instances the clerk of the peace who acted as clerk to the
justices also acts as clerk to the county council. By the
Burgh Police (Scotland) Act, 1892, "police burghs" are
now for the first time constituted separate districts under
the Weights and Measures Acts, the magistrates of a burgh
being the local authority, and certain powers are given to
the Burgh Commissioners of Police with the object of
preventing fraud in the sale of bread and coal or in the
weight or quantity of grain, hay, or straw, or other
commodities brought for sale within a burgh. Power
also is given to the chief constable, or any other constable

specially appointed, or any inspector of weights and measures within a burgh, to enter premises and to require articles to be weighed. A "police burgh" is any populous place in Scotland containing 700 inhabitants or upwards.

In Scotland there are 87 inspectors of weights and measures; or in Great Britain there are 521 inspectors, 314 county officers, and 207 borough officers.

In Ireland, excepting in the City of Dublin, and the Dublin Townships of Blackrock, Dalkey, Kingstown, New Kilmainham, Pembroke, Rathmines and Rathgar, the inspectors are appointed by the constabulary authorities, such head or other constable in each petty session district including the Cities of Belfast, Cork, Limerick, and Londonderry, and the towns of Galway and Sligo, being selected by the Inspector - General of the Royal Irish Constabulary as ex-officio inspectors. These ex-officio inspectors have to obtain a certificate of qualification from the Board of Trade before they can act as inspectors. *Local practice in Ireland.*

In Dublin the inspector is appointed by the Corporation; and in the Dublin Townships by the Township Commissioners. In former times the Lord Mayor of Dublin himself visited the public markets, and with his officers inspected traders' scales and weights.

The Inspector-General of the Royal Irish Constabulary issued in 1891 regulations for ex - officio inspectors of weights and measures in Ireland with respect to the verification and inspection of weights and measures and weighing instruments (Messrs. Hodges, Figgis, & Co., 104, Grafton Street, Dublin); and a description of the standards and instrumental equipment to be provided by local authorities in Ireland was also issued in 1892.

In Ireland there are outside the City of Dublin, and the adjacent Dublin Townships, 573 ex-officio inspectors, one for each petty session district, the number of such districts in each county varying with the size of the county.

In the United Kingdom, there are at present 1,099 inspectors.

Denomination of Board of Trade Standards. The Weights and Measures Act, 1878, s. 24, provides that every person who uses or has in his possession for use for trade a weight or measure which is not of the denomination of some Board of Trade standard, shall be liable to a fine not exceeding five pounds, or in the case of a second offence ten pounds and the weight or measure shall be liable to be forfeited. A list of Board of Trade standards will be found in Appendix 1.

The same Act also provides (s. 29) that every measure and weight whatsoever used for trade shall be verified and stamped by an inspector with a stamp of verification under this Act. On weights, measures, and weighing instruments used in trade, the inspectors now place a stamp of the design included in the ring—a crown, sovereign's initials, and the number of the officer's district :—

INSPECTORS' STAMP.

The Weights and Measures Act, 1889 (s. 1) provides that every weighing instrument used for trade shall be stamped.

Board of Trade "Model Regulations." The latter Act also provides (s. 9) that every local authority within the meaning of this Act shall, with the approval of the Board of Trade, make for the guidance of the inspector appointed or employed by that authority or person, general regulations with respect to—

(*a*) the procedure to be observed in the verification and stamping of weights, measures, and weighing and measuring instruments, including the prohibition of stamping in cases where the material or mode of construction appears likely to facilitate the commission of fraud ; and

(*b*) the inspection of weights and measures and weighing and measuring instruments.

With the view of giving effect to this provision the Board of Trade issued, in 1890, " Model Regulations" for the use of local authorities, in which modes of verifying and stamping weights and measures and weighing and measuring instruments were set out.

Although at present the inspection of weights and measures in rural districts is carried out with nearly as much accuracy as it is in urban districts, yet formerly there was practically no local administration of the weights and measures laws in country districts, but only in the cities and boroughs. Some local verification of weights and measures has been attempted, however, in England from the earliest times ; by a Statute of William the Conqueror, it was ordered that the weights and measures should be true, and stamped in all parts of the kingdom.

Ancient local practice.

King Richard, in 1197, directed that the custody of the assize, or of the King's Standards, should be committed to certain persons in every city and borough ; and by the Statute 51 Hen. 3, Stat. 6 (1266), it was ordained that " Six lawful men shall be sworn truly to gather all " measures of the town." In a statute of Edward I., it was also ordained that " the standard of bushels, gallons, " and ells, shall be sealed with an iron seal of Our Lord " the King, and no measure shall be in any town unless it " do agree with the King's measure."

By a Commission, under an Act of Edward III. (1340), two good and sufficient persons in every county were required to be appointed to survey all weights and measures; practically, however, this Commission was confined to the towns, and when it expired in the time of Richard II. its duties became part of the ordinary business of the magistrates of cities and boroughs.

In 1495 more detailed regulations were established by the Act 11 Henry 7. c. 4; the head officers of every city and town were to cause, twice in the year or oftener, all weights and measures to be examined, such as were found to be defective were to be immediately destroyed and the parties fined, and for further punishment, after the third offence, " to be set in the pillory, to the example of others."

The earlier statutes sometimes provided against the use in trade of particular kinds of instruments; thus an Act of 25 Edward 3. c. 9 (1351) provided that " auncel " weight should " be wholly put out, and that every person " do buy and sell by the balance, so that the balance be " even." The auncel instrument was used in a manner somewhat similar to that of the steelyard, but scales or hooks were hung at each end of the beam, which was lifted in the middle with the finger and hand. The statute 27 Edward 3. Stat. 2. c. 10 (1353) also enacted that every kind of " avoir du pois " be weighed by equal balances, and without inclining to one side or the other. In 1429 another Act was passed abolishing also auncel weight, and providing that every city and town should have a common balance with standard weights, at which the inhabitants might freely weigh without paying anything.

The control of weights and measures in Ireland was not always carried out so well as it is now, for a commissioner appointed by the Lord Lieutenant in 1852 to inquire into the state and condition of the fairs and markets of the country recommended legislation of a drastic character for the weighing of produce brought into the market and for the assimilation of weights and measures. In Ireland, where the use of market weighing machines is common, the weighing fees still form a considerable portion of the market revenue. The special legislation with reference to the mode of weighing in markets (Weights and Measures Act, 1878, s. 77), in many instances appears to have become necessary in consequence of the vendor being deprived by the market authorities of a portion of his price, either by not getting paid for the full weight of his goods, or by something being deducted from the price as an allowance from the actual weight, or in the name of porterage or storage, or as house money, or for signing a ticket.

In each town the King's Officer, who saw that all weights and measures were "answerable" to the standards, was formerly known as the " Clerk of the Market," and his powers and duties are particularly prescribed by two ancient statutes of Edward III. (1340), and Charles I. (16 Chas. 1. c. 19). One of the latest recorded instances of a Clerk of the Market exercising powers as a King's Officer was in Middlesex, in 1738, when the Clerk of the Market of the King's Household was authorised by patent to inspect all weights and measures within the precincts of the King's Palace or the "little virge." Market practice.

Closely connected with the right of holding a "market" was that of keeping standard weights and measures, and market owners in some cases formerly provided sworn meters for measuring cloth, corn, salt, &c. It was the duty of the owners of markets to examine weights and measures, and to hold assize of cloth and of bread and ale.

The Markets and Fairs Clauses Act, 10 and 11 Vict. cap. 14, 1847, contains provisions with reference to the weighing of goods and carts in markets. Market authorities or undertakers are to provide sufficient and proper weighing houses or places for weighing or measuring commodities sold in the market or fair; and to keep proper weights, scales, and measures according to the standard weights, &c. for the time being. The market authority are also to provide buildings or places in which to weigh carts, and to keep in those places machines and weights proper for the purpose. The provisions of the Act of 1847, so far as they relate to public markets, are incorporated with the Public Health Act, 1875; with the Public Health, Ireland, Act, 1878, and with the Burgh Police, Scotland, Act (Royal Commission on Market Tolls and Rights. Final Report, 1891, p. 164), so that urban sanitary authorities may have the powers given by those provisions for establishing or regulating markets, with respect to the weighing of goods and carts.

The Report of the Royal Commission above referred to shows that the modes of weighing, charges for weighing, and market deductions from ascertained weighings, vary according to locality and custom.

Courts Lect. In many country districts "Juries" also were nominated, mostly by the Lord of the 'Manor, at a Court-Lect (leta visus francii plegii), for the purpose of inspecting weights and measures; and by a statute of 1325, juries were directed to present " such as have double measure, and buy by the great and sell by the less." The Courts-Lect are said to be the most ancient courts of the land, and they were held either under the King's Charter or by ancient prescription and usage. As a typical instance of the extent of these ancient manorial jurisdictions, reference may here be made to the manor of Wakefield, which had an area of nearly 148,000 acres, with a population of 325,000; the judge of the Courts-Lect, at which offences were heard, being the steward of the manor. The jury was made up of representatives from every township within the bailiwick. When the jury were satisfied of the commission of any offence against the weights and measures laws the steward fixed the amount of fine, two of the jurors, called afferors, having power to reduce the fine if they thought there were extenuating circumstances. Previously to the reign of Charles I. this manor was in abeyance of the Crown, it having come into the hands of Edward IV. as the heir of the Duke of York, who was the original lord of the manor. In 1892 the County Council of the West Riding purchased the franchise of the manor an Act of Parliament (55 & 56 Vict. Ch. 18) having been passed in that year to empower County and Borough Councils, by agreement with the manorial authority, to purchase franchises of weights and measures.

Manorial jurisdiction was largely exercised in England up to the year 1835, but now there are probably not a dozen courts leet who appoint inspectors.

During the last century in a few cities and towns, particularly in London (as in Westminster), "Annoyance juries," or inquests, were appointed by the vestries, and such juries had similar powers of presentment and inspection as were formerly exercised by the juries appointed at courts leet. By the Act 29 Geo. 2. c. 25 (1756), such juries had powers to seize and break-up false scales, &c., but their powers practically ceased on the passing of the Weights and Measures Act, 1878. (See also Act 24 & 25 Vict. c. 78. 1861).

The Merchandise Marks Act, 1887 (50 & 51 Vict. c. 28), makes it an offence to apply any false "trade description" to goods and enacts that the expression "trade description" means any description, statement, or other indication, direct or indirect, as to the number, quantity, measure, gauge, or weight of any goods.

Besides the local authorities already named, certain trade guilds formerly exercised their powers under royal charters to verify and inspect weights and measures used in their particular crafts.

E 2

Gold-
smiths'
Company.

Of such guilds, those in the City of London were
the more important; the Goldsmiths' Company, for
instance, whose charters extend back to Edward III.
(1327) and Richard II. (1392), at one time exercised
supervision over the weights used by goldsmiths, and
all troy weights so used in London were sealed at
Goldsmiths' Hall up to the year 1679. During the
restoration of the standards in 1588, the Goldsmiths'
Company gave valuable assistance to the Royal Commission,
who had to construct the new standards of weight; and
we find in the Report of the Commission that a set of the
new standard troy weights was delivered to the " London
Societat aurifabyr." which set of standards appears,
however, to have been destroyed at the Great Fire of
London in 1666.

The Goldsmiths' Company conduct the proceedings of
testing the new coin of the realm at the annual Trials of
the Pyx, which are held at Goldsmiths' Hall, Foster Lane,
London.

Founders'
Company.

The charter of the Founders' Company in London,
granted by King James in 1614, also gave that company
the right of stamping and adjusting all avoirdupois weights
sold or used in London, and weights marked with the
mark of the Founders' Company were declared to be legal
for use throughout the United Kingdom.

All brass weights sold or used in London were then
required to be twice stamped, once by the Founders'
Company, and once by the stamper appointed by the
Corporation of London; but by the passing of the Weights
and Measures Act, 1889, a person using weights in London
is not required to have them stamped more than once
now.

Coopers'
Company.

By a statute of 1531 the Wardens of the Coopers
Company were authorised to search, view, and gauge,
within the City of London and suburbs, all manner of

barrels, or other vessels, used or sold for ale or beer, **and to mark such as agree in their contents** with **the mark of St. Anthony's Cross.** In 1589 **the** Company were **also** required **by statute (31** Eliz. c. 8) to visit any beerhouse in London, or within two miles, to gauge and mark the casks. In 1660 it was enacted by 12 Car. 2. c. 23. that every **36** gallons of beer was to be reckoned a barrel, and **every 32** gallons of ale in like manner reckoned a barrel of ale, thus confirming the provisions of **23 Henry 8. c. 4,** under which beer, and also barrels **containing such number of** gallons respectively, **were to be verified by the Coopers'** Company. The **company do not appear to have carried** out their duties as to **testing casks** after the year **1758.**

The Plumbers' Company of London under a **charter** from James I. (1611) **were** authorised to search for **and to** try all weights **made** of lead. Such search **is** now no longer needed, as it is illegal to use **in trade** a weight made wholly of lead or pewter. Plumbers' Company.

The Fruiterers' Company **of London** also had **power to** seal measures **used** in the sale **of** fruit; powers no doubt which **at** one time **were usefully exercised.** In **1702 the** Act 1 Anne c. 15 was passed **to ascertain the** "water measure of fruit" (a round measure $18\frac{1}{2}$ inches in diameter **and 8 inches** deep), but **this Act** did not extend to **measures** sealed and allowed by the Fruiterers' Company. Fruiterers' Company.

The Merchant Taylors' Company formerly tested lineal measures and the lengths of cloths. The beadle of the company attended annually **at St.** Bartholomew's Fair, London, up to the time of **its** discontinuance, **and** there tested the measures, **&c.** The company possess **an** ancient silver standard yard cloth measure divided into nails (1 nail **=** 4 inches **or** 4 thumbs), constructed at some time anterior to the year 1585, and when compared with the present Imperial yard **in** 1892 it was only found **to be 0·004 inch** longer. **The** measurement of cloth was Merchant Taylors' Company.

formerly regulated by public statute, thus the aulnage of manufactured cloths was ordained by 2 Ed. 3. c. 14 in 1328, and when found correct the cloth was to be marked by the King's bailiffs.

Weights and measures in docks and harbours.

In some large docks the Harbour Board appoint a separate officer to test their weights, &c., and then such officer is sometimes responsible to that board for the supply of accurate weights and weighing instruments required for weighing goods discharged from vessels lying in the docks. In the case of the Mersey Docks and Harbour Board the annual cost of the "Weights and Measures Department" of the Harbour Boards amounts to 3,000*l.*, and the "superintendent of weighing materials," is stated to test annually no less than 20,000 weights, 900 scale beams, and above 100 large weighbridges. As to the inspection or examination of weights and measures for other special purposes, *see* Part VI.

Standards Commission, 1868-70.

The inspection and verification of weights and measures was reported on at considerable length by the Standards Commission (*see* p. 5), who in their Fourth Report (21 May, 1870) recommended many improvements of the then existing system of local inspection. The Commission made five Reports, dated respectively 24 July, 1868, 3 April, 1869, 1 February, 1870, 21 May, 1870, and 3 August, 1870. Their first report was merely of a preliminary character; the second report dealt with the Metric System, recommending its permissive use in trade; the third report recommended the abolition of Troy Weight; and their fifth and final report referred to the Standards Department, particularly to the steps taken by the Commission for promoting and maintaining the efficiency of the Board of Trade Standards of measure and weight. The Chairman of the Commission was Sir G. B. Airy, K.C.B., the Astronomer Royal.

Under the Weights and Measures Acts the following Orders in Council have come into force.

1870, Aug. 9.	Legalising	new standard coin weights (including the weights of £5 to ½ sovereign).
,,	,,	new standard decimal grain weights.
1871, Mar. 24.	,,	new liquid measures of capacity, gas measures, and measures of length (with subdivisions).
,,	,,	errors to be allowed in comparison of copies of the official standard weights and measures.
1876, June 27.	Legalising	measures of 100 feet and of the chain of 66 feet with subdivisions laid down in Trafalgar Square as secondary standards.
1879, Feb. 4.	,,	**new scale of** errors to be tolerated in local **standards.**
,,	,,	**cental** or new hundredweight.
1879, Aug. 14.	,,	**new** standards of apothecaries' **weights and** measures.
1880, Feb. 26.	,,	4-bushel measure.
1880, April 28.	,,	5-gallon measure.
1881, May 18.	,,	additional denominations of standards.
1881, Aug. 26.	,,	Whitworth's external and internal cylindrical and external plane gauges.
1882, June 29.	,,	new scale of errors to be **tolerated on local** standards.
1883, Aug. 23.	,,	new denominations of standards (wire gauge).
1886, Aug. 3.	Approving	the use of new Parliamentary copies of the Imperial Standard Yard and the Imperial Standard Pound.
1889, Nov. 28.	Legalising	**measures of** capacity of 6 **to 31** gallons inclusive.
1890, Mar. 21.	Fixing	fees to be taken by inspectors on the verification of liquid measures from 32 gallons to 1 gallon inclusive.
1890, May 1.	Exempting	borough of Nottingham from provisions (sub-section (a)) of the **27th section** of the Weights and Measures Act, 1889.
1890, Nov. 22.	Legalising	measure of capacity of 3 gallons.
1893, Mar. 15.	,,	scale of errors to be tolerated on local standards when locally re-verified by a comparison in the presence of a justice of the peace.
1894, Aug. 23.	,,	new denominations of standards for the measurement of electricity.

No account **of** our **weights** and measures would be complete without some reference to the weighing instrument by which differences of weight are determined. The use, however, of the weighing instrument is so extensive and varied—from the delicate assay balance to the heavy railway weighbridge—that a volume might alone well be filled with a description of the different kinds of weighing instruments and modes of weighing. English literature at present is not rich in books on weighing instruments generally, and until the issue of the Board of Trade " Model Regulations " in 1890 there had not been made any attempt (with the exception of useful trade catalogues published by our principal manufacturers of scales and weighing machines) to classify the different kinds of weighing machines in use for trade. There is perhaps, still wanting an exhaustive English treatise on the construction of weighing instruments, similar, for instance, to the German work of Ernst Brauer (Weimar, 1887). Mr. Wilfrid Airy, C.E., however, in 1892, read before the Institution of Civil Engineers a paper on weighing machines (Proceedings, Institution of Civil Engineers, Vol. CVIII. Session 1891–92. Part II.) which contains diagrams of the forms of the principal weighing-instruments now used in trade, and permission has been kindly granted by the Institution to reproduce here (*see* Appendix No. 2) these diagrams of representative types of weighing instruments.

The " Model Regulations " above referred to give the following "general description " of weighing instruments:

The ordinary **weighbridge** or weighing-machine is a *compound lever machine* with unequal arms for increasing the power of weighing without the equivalent use of standard weights, the **weight being indicated by an attached steelyard** or on a dial. In its portable form such machine is known as the " platform-machine," and in its largest form, when fixed for use in weighing carts, &c., it is known as a " weighbridge."

In the *spring* platform weighing machine there are no levers excepting those of a simple form on which a platform rests, but the **weight** or load is balanced by means of springs, and is recorded on a dial.

The weighing **machine does** not therefore include any scale, balance, steelyard, or **other machine or instrument** for weighing with which it is necessary to use standard weights, **or in which only one** lever or beam is used.

The different kinds of **weighing machines are as follows :—**

Weighbridges, as cart, **waggon, truck, or railway weighbridges :**

Compound lever **weighing machines:** These include dormant machines, or those whose platforms are level with the ground; and portable **machines,** with or without wheels, **as** used at railway stations, &c.

Suspended or crane weighing machines :

Automatic weighing machines :

Spring platform weighing machines.

General description of Scale-beams and portable Weighing **Instruments.**

Simple lever instruments with equal arms include : —

Any balance which is in equilibrium when the **weights in the two pans** are equal ;

Assay, bullion, or chemical balances ;

Box end and Dutch end scale-beams ;

Counter, or **stand-scales ;**

All equal-armed instruments where the pans are suspended **from the** beam or lever ;

Counter or inverted weighing machines, with more than three bearings ;

Accelerating machines ;

Vibrating machines ;

Equal balanced coal machines ;

Any instrument with equal arms and with simple levers, **where the pans** or platforms are placed over the levers.

Simple lever **instruments with unequal arms include :—**

Roman steelyards **with a** single **graduation, traversing** poise, and suspended hook or pan ;

Roman steelyards with more than one graduation, or traversing poise ;

Letter balances ;

Any instrument with unequal arms where the pans or platform are suspended from the beam or lever ;

Portable coal machines ;

Counter steelyards, for manufacturers' use only ;

Automatic weighing machines ;

Coin balances ;

Any instrument with unequal **arms and with** simple levers, **where the** pans or platforms are placed over the levers.

Spring balances include :—

Any weighing instrument when the weight **or load is balanced or deter**mined by means of springs ;

Weighing instruments where the weight or load is placed under the spring;

Weighing instruments where the weight or load is placed over the spring;

Any instrument where the difference of weight is determined, otherwise than by means of a lever.

Inspector's weighing instruments.

A form of scale-beam used by inspectors of weights and measures is shown in Fig. 14. The beam is made of gun metal, polished bright, in four sizes for indoor use and three sizes for outdoor use, and it has the dimensions between the two " knife-edges " as shown in Fig. 13. For indoor use the thickness of the beam may be about $\frac{1}{60}$th of its length, the depth near to the centre bearing being about $\frac{1}{12}$th of the length of the beam, the depth at the ends being about one half of the depth at the centre. The beam may or may not, as the inspector thinks fit, be furnished with means for adjusting it for daily wear.

Beam marked to weigh.	Length of the Beam between the two end Knife-edges. For Indoor Use.	Beam marked to weigh.	Length of the Beam between the two end Knife-edges. For Outdoor Use.	
56 lb.	28 inches.	56 lb.	24 ins.	In these beams the depth near to the centre should be about one-eighth of the length.
7 ,,	16 ,,	7 lb.	14 ins.	
1 ,,	10 ,,	1 lb. and under.	8 ins.	
1 oz.	6 ,,			

This form of beam has open and continuous bearings, and is of such dimensions as to secure during the weighings sufficient sensitiveness and stability with the least amount of metal in the beam when carrying its maximum load. The 56-lb. beam will turn with 5 grains, the 7 lb. with 1 grain, the 1 lb. beam with 0·2 grain and the 1 ounce with 0·01 grain.

INDOOR USE. (*a, b, c, d*).

OUTDOOR USE. (*e, f, g*).

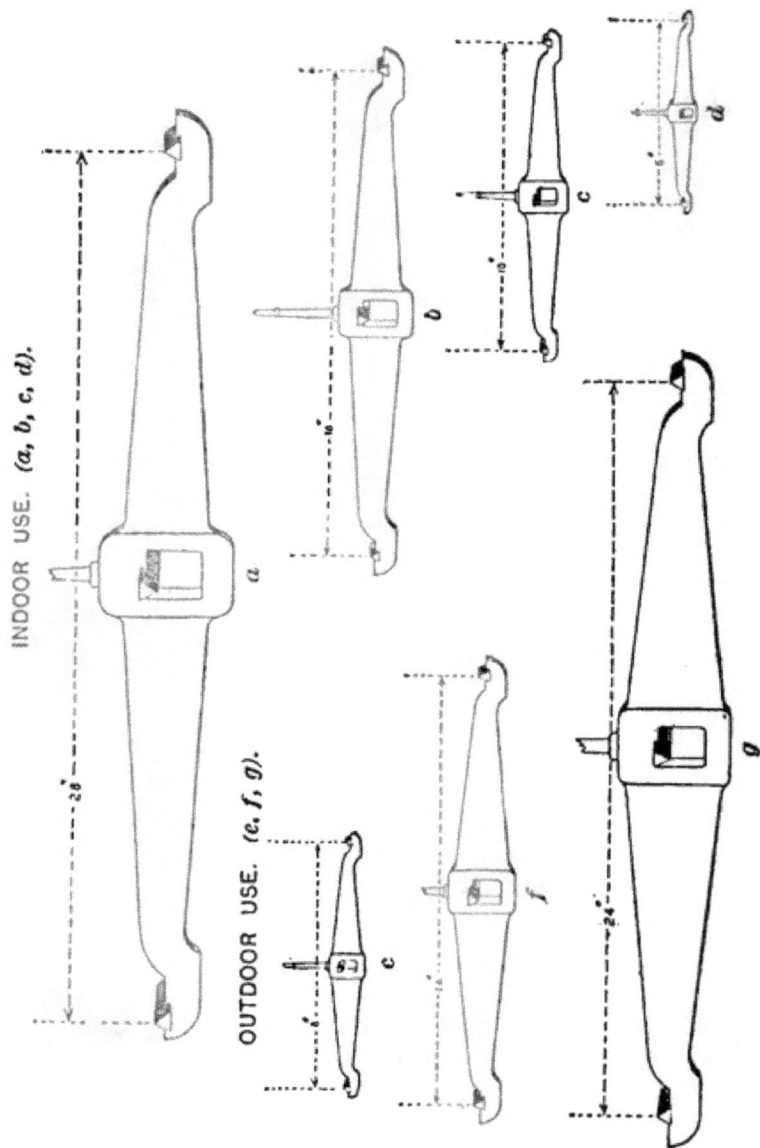

Fig. 13.—INSPECTOR'S SCALE-BEAMS, SHOWING DIMENSIONS.

Fig. 14.—ONE FORM OF AN INSPECTOR'S SCALE-BEAM, WITH CHAINS
AND PANS COMPLETE, AND SUPPORTED ON A "STAND."

PART III.

————◦|◦———

12. MEASURES USED FOR SURVEYING PURPOSES OR FOR THE MEASUREMENT OF LAND.

In a trigonometrical survey like that of the Ordnance Trigono-
Survey Department of Great Britain (now a department metrical
of the Board of Agriculture), the object of the survey may surveys.
be said to be to determine the position of various points
on the earth's surface which are far apart one from the
other, and consequently to measure the distance between
such points, and our cadastral maps are the outcome of
such surveying. Geodetical measurements, on the other
hand, relate mainly to the mathematical determination
of the general figure of the earth.

For trigonometrical purposes there are measured "base-
lines," of from 3 to 10 miles in length, by means of
standard bars, made of iron or steel, of the length of
10 feet or of 1 or more metres; and from such base lines
there starts a system of triangulations by means of which
the positions of the various points on the earth's surface
are determined. An account of the Great Trigonometrical
Survey of this country begun in 1783 has been given by
General Sir H. James, R.E. (London, 1858); and accounts
of other trigonometrical surveys abroad have also been

published, as of the Survey of India by General J. T. Walker (Dehra Doon, 1870); "Base Centrale de la Triangulation Géodésique d'Espagne" (Madrid, 1865); "The Geodetic Survey of the United States" (Washington, 1866–1894); Report on "The Geodetic Survey of South Africa," by Dr. Gill (Government Printers, Cape Town, 1896,), &c.

The principal base line measured in Great Britain was originally that on Salisbury Plain, by General Mudge in 1794; and in Ireland that of Lough Foyle in 1827–8, by General Colby; and a plan of the principal triangles in England and Wales and part of Scotland, showing the connections with the base lines on Salisbury Plain and Lough Foyle, has been issued by the Ordnance Survey Department. The difference in the British measurement of the Salisbury Plain base (1849) from that obtained by calculation through a large number of triangles starting from the Lough Foyle base was not more than 0·2 feet. The most recent of direct trigonometrical measurements is perhaps that of South Africa, when a base line of 6,000 feet was measured in two sections by means of standard bars, the probable sum of errors of one section being stated to be only ± 0·023 inch and of the other section ± 0·016 inch. All the English measurements are referred to the Imperial yard, a copy of which is kept at the headquarters of the Ordnance Survey at Southampton, the normal temperature followed being that of the Imperial Yard, viz., 62° Fahrenheit. An important statement as to the progress of the Ordnance Survey to 31st December 1893, may be found in the recent Report of a Departmental Committee, which included Sir J. Dorrington, Bart., M.P.; Sir A. Geikie, F.R.S.; Lieut.-General Cooke, C.B.; and Major D. A. Johnston, R.E., as Secretary.

As the length of a metal bar (as a 10 ft. steel survey standard) depends on the temperature of the bar, and as trigonometrical measurements have to be made in different latitudes, and at various temperatures, it becomes necessary to allow for the expansion of a bar by heat. For such purpose the rate of expansion of a bar may be experimentally determined (for instance, it has been found that a 10 ft. steel bar expands 0·0007632 inch for a rise in temperature of 1° Fahrenheit, or its "coefficient" of linear expansion for 1° Fahrenheit would be represented by 0·00000636). Or the expansion of a bar may be allowed for mechanically, as in the English and Indian surveys, by the use of compensating bars.

In European surveys the old French toise and the old Austrian klafter were formerly used; now, however, the most of European geodedic surveys are expressed in metres :—

$$1 \text{ metre } = 1·09361426 \text{ yard.}$$
$$1 \text{ toise } = 2·13151116 \text{ ,,}$$
$$1 \text{ klafter } = 2·07403483 \text{ ,,}$$

The earliest triangulations, as those of Picard in France in 1667, which were referred to the toise, are stated to have enabled Sir Isaac Newton to establish finally his doctrine of gravitation, as published in his Principia in 1687, when he showed that the earth must be in the form of an oblate spheroid, and that gravity must be less at the equator than at the poles, as was proved by subsequent measures with a toise standard in Peru in 1735.

For field and office work a variety of survey units are used with a corresponding diversity of scale. The maps of our Ordnance Survey are set out in Imperial units as follows :— *(Measurement of land.)*

$\frac{1}{500}$ or 126·720 inches to the mile.
$\frac{1}{2500}$ or 25·344 inches to the mile.
$\frac{1}{10560}$ or 6 inches to the mile.
$\frac{1}{63360}$ or 1 inch to the mile.

The $\frac{1}{2500}$ scale maps cover almost the whole of the towns and parishes, and the 6-inch maps the whole of the United Kingdom. Town maps are also published to the scale of 5 feet and 10 feet to the mile. (*See* Catalogue of Maps, issued by the Director-General, Ordnance Surveys, 1896.)

For plotting land measurements surveyors use finely divided scales showing 20 chains to a quarter mile, and 2 chains to the inch, represented by 200 divisions to the inch. In the mensuration of plane surfaces the following table of measures is followed :—

62·7264 square inches = 1 square link.
625 square links   = 1 square pole or perch or rod.
10,000 square links   = 1 square chain.
25,000 square links   = 1 rood.
10 square chains   = 1 acre.
100,000 square links = 1 acre.

As one link equals 7·92, inches we may find the area of a rectangular piece of ground in the following manner : Suppose the length of the ground is 792 links and its breadth 385 links, find its area in acres, roods, and perches. 792 × 385 links = 304,920 square links; as 40 perches equal 1 rood and 4 roods equal 1 acre, cut off five figures from 304,920 square links and the result will be 3·04920 acres. Multiply the decimal 0·04920 by 4 and 40 and the result will be—Answer = 3 acres, 0 roods, 7·872 perches.

In the United States the Imperial units appear also to be followed in private practice, drawings being made of so many feet to the inch, and where a surveyor uses a Gunter's chain of 66 feet, he plots his work to so many chains to the inch.

Engineers in using a 100 feet chain, and measuring rods divided into decimal parts of a foot, also generally reduce maps to the inch units.

A measuring **wheel** (viameter) is used for measuring Portable approximately distances along roads, as cab distances, &c. land The wheel **is rolled over** the ground to be measured **and** instru- the motion is communicated to a series of toothed wheels ments. so proportioned that the index wheel registers the revolutions of the measuring wheel. Of course unless a straight or given line is followed, this measuring **wheel will** give uncertain results. Various other forms **of instruments are** also used for such purposes **known as odometers, distance** indicators, way-measurers, **&c. The pedometer is an** instrument sometimes **used for counting and** recording the number of steps taken by **a person carrying it.**

Other **and more accurate** instruments are also **used in** the survey of **land, as telemeters for** finding the **distances** of inaccessible **objects;** or **for** measuring and calculating the areas of **plane** surfaces, as planimeters. Of course **in** the measurement of " base-lines," **for the determination of** astronomical positions other well known instruments **are** used, as the theodolite, **zenith sector or transit instrument,** &c.

In measuring lines **on** maps, whether curved or straight, without the use of a rule or linear scale, **small** instruments **are** sometimes used known as the diagrammeter, opisometer graphometer, &c.

Steel chains of 100 feet and of 66 feet divided into Chain links, were at one time used in the Ordnance Survey, and measure- were stretched in wooden-trays between **a** drawing-post ments. and a weight-post with a " pull " of 56 lbs.

In India the earlier survey measurements of Colonel Lambton **up** to 1826 were made by a Ramsden 100 feet steel chain, of **a form** described in the Philosophical Transactions of the **Royal Society. Chain** measures are, of course, still commonly used in **the** Ordnance Survey, and chaining officers have the **following** instruction :—

" The chain and tape (when it has to be used) must be compared with the standard every morning on parade, and

F

any error must be corrected before commencing work. All chaining must be horizontal (on $\frac{1}{2500}$ and $\frac{1}{500}$ scales) and not along the surface of sloping ground. In chaining on sloping ground too long a portion of the chain should not be used at once. In chaining lines the chain must be extended, but neither stretched too tightly nor allowed to rest too slackly. A line with an error of two links in every 1,000 will be passed."

The land surveyors' chain known as Gunter's chain, of 66 feet, is divided into 100 "links," each link being therefore 7·92 inches, or 1 foot of chain = 1·515 link. This chain of 66 feet is a convenient measure of superficies as by it 1 acre is equal to 10 square chains or 100,000 square links. Professor Edmund Gunter, who invented the chain of 66 feet about the year 1620, also invented, in 1624, the " line of numbers," and he introduced several improvements into surveying instruments during the time he was connected with Gresham College in the City of London.

The old Scotch acre was also divided into ten square chains, each chain into 16 "falls," and the whole into 100,000 square links.

Engineers also use iron or steel chains of 50 or 100 feet, divided into links, each iron or steel link with its connecting ring being 1 foot long, having their true lengths at the temperature of 62° Fahr. Although chain measures are usually made of iron or steel, those used in mines are often made of brass. In metalliferous mines (Brough's treatise on mine surveying, 1889) a chain of 10 fathoms, or 60 feet, is stated to be used, each link being 6 inches long, the chain being provided with brass marks at every fathom. It would appear that for colliery use the chain may be left half an inch longer than the true length since it is rarely drawn out into a straight line. In continental countries the metric chain is used for mining, but a unit somewhat similar in length to the " fathom " is also used. The fathom is declared by the Weights and Measures Act, 1878, to be equal to 2 yards, or 1·8288 metres.

Deal rods and glass rods were subsequently substituted for chains, and in the measurement of **the** base line on Hounslow Heath in 1784 the difference obtained between the measurements by glass rods and the more convenient steel chains was only 2¾ inches in 27,404 feet. **Rods made** of deal, pine, or lance wood, are used **in surveying and in** checking chain measurements; such rods are **usually 5, 6,** or 10 feet in length, and mostly have **their ends protected** by brass plates, after being soaked in **boiling oil and well** protected **by** varnish, so as **to** lessen the effect **of humidity.** Even, however, with the utmost care rod measurements can hardly be relied on beyond 0·03 inch in 100 feet.

In Russia base lines of over 2 miles in length **have** been measured with steel wires with an accuracy **of** 0·1 inch in the whole length of the line.

Measurements can be made more accurately **by steel** ribands or steel wires than by linked chains, a **measure-** ment by riband may be taken **as one and a half times** more accurate than a chain **measurement. For all** surveying purposes a steel riband of 50 feet should have a width of not less than ½ inch and be about ₃₂¹-inch thick. Of course much depends on the mode in which a chain measure or a steel riband may be used; for we have not only to ascertain the absolute length of **a chain** in yards or metres, but **we** must allow for the " sag " of the chain, for which latter purpose the modulus of elasticity of a steel chain may be found experimentally by applying to it varying stretching weights or " pulls." At the Standards Office a pull or normal tension of 56 lbs. is given to steel 100 feet chains; 40 lbs. on steel 66 feet chains; and 10 lbs. on steel ribands. The distance indicated **by a** 500 feet **steel** riband when subjected to

Measurements by ribands.

a normal tension has been found to be 500·082 feet. On
the best wire-woven linen tapes a normal tension of 12 lbs.
is given, on ordinary linen tapes the tensions are 5 lb. on
100 feet, 4 lb. on 66 feet, 2½ lbs. on 30 feet linen tapes.
A wire-woven 100 feet linen tape will stretch nearly
¾-inch with a 12 lb. pull.

In the measurement of ships the Board of Trade
"Instructions and Regulations" (1895) permit the use to
some extent of the measuring tapes, but all tapes are
required to be frequently tested as to their accuracy.   As
all flaxen and hempen manufactures are liable to contrac-
tion by moisture no tapes are to be used which are not
waterproof ; and those only of such length as (15 feet,
60 feet) which involve no practical error arising from
"sag" or deflexion, or from extension by long continued use,
and it is also stated that a strong waterproof tape of
100 feet long, decimally divided into feet, held moderately
tight, will be found convenient in measuring the lengths of
vessels, and the breadths of the areas, and in measuring
the lengths and breadths of closed-in spaces.

Mural and   Of whatever material a chain or tape may be made it
reference
Standards. requires to have its accuracy periodically verified by a
fixed mural or reference standard, and hence long standards,
or small base lines of 100 feet and 66 feet, with their
subdivisions, including the land measure of the pole or
perch (5½ yards) have been laid down by the Board of
Trade on the north side of Trafalgar Square, London, and
by different local authorities throughout the United
Kingdom.   The standards at Trafalgar Square were made
legal measures by an Order in Council dated 27th June
1876, and copies of these standards were subsequently laid
down at the Arts Museum, Edinburgh, the Guildhall,
London, the City Hall, Dublin, the Assize Courts, Manchester, the Municipal Buildings, Glasgow, the Town Hall,
Bradford, &c.

Besides these public standard chain measures, bronze mural tablets are also exhibited at the places above referred to, as well as on the outside of the Royal Observatory, Greenwich, (the original mural standards), and in the vestibule of the Science Galleries at South Kensington Museum. These tablets show the length of the yard, foot, and inch (and at Kensington and Dublin of the metre, decimetre, and millimetre also), so that any workman can at once test his 2-foot rule to one hundredth of an inch.

The connection between our system of weights and measures and pendulum measurements is no longer a direct one, for the yard measure is no longer based on the length of the seconds' pendulum. Pendulum observations, however, form an important part of geodetic surveys, as those in India carried out by Capt. J. P. Basevi, Col. W. J. Heaviside, and Col. Herschel, or those carried out at the United States Geodetic Survey by Mr. Peirce ; and by M. Sawitsch in Russia. The object of pendulum observations may be said to be to determine the figure of the earth, as the length of the seconds' pendulum varies with the latitude at which the pendulum may be swung. The ratio of the earth's axes, as deduced from pendulum observations, is, according to Clarke ("Geodesy," Oxford, 1880), as 292 to 293.

In the **C. G. S.** (centimetre—gram—second) system of units the following values may be taken for the apparent acceleration of a body falling freely under the action of gravity in vacuo (g), and for the length (l) of the seconds pendulum connected with the value of (g) :—

(margin note: Pendulum measurements.)

			g	l
Equator	-	-	978·10	99·103
Paris	-	-	980·94	99·390
Greenwich	-	-	981·17	99·413
Berlin	-	-	981·25	99·422
Pole	-	-	983·11	99·610

PART IV.

13.—SCIENTIFIC MEASURING INSTRUMENTS.

In the measurement of physical quantities the units of such quantities may be classified as follows :—

>*Fundamental* units, as those of length, mass, and time.
>
>*Geometric* units, as those of surface, volume, and angle.
>
>*Mechanical* units, as those of velocity, acceleration, force, work, power, pressure, and moment of inertia.

With reference to the units of length and mass, standards are determined, to which reference has been already made, and in connection with other units various forms of measuring instruments become necessary, of which, in some instances, the Legislature has taken cognizance, as in the measurement of electricity, gas, &c., and to these some reference may be offered here. (*See also* Guillaume's "Unités et Étalons" (Paris, 1890), Everett's " Illustrations of the C.G.S. System of Units " (Macmillan, 1891), Lupton's " Numerical Tables " (Macmillan, 1892).

In the measurement of electricity the instruments are standardised by reference to verified standards, and standards and instruments are deposited with the Electrical Adviser of the Board of Trade, Major P. Cardew, R.E., at the Laboratory in Richmond Terrace, Whitehall, as follows :—

CURRENT MEASURING INSTRUMENTS.
Standard Ampère.

Sub-standards for measurement of currents as follows :

1 to	5	ampères.	
5 ,,	30	,,	
30 ,,	120	,,	
100 ,,	600	,,	
500 ,,	2,500	,,	

POTENTIAL MEASURING INSTRUMENTS.
Standard Volt.

Sub-standards for measurement of pressures as follows :

25 to	200	volts.
50 ,,	400	,,
100 ,,	800	,,
200 ,,	1,600	,,
400 ,,	3,000	,,

RESISTANCE MEASURING INSTRUMENTS.
Standard Ohm.

Sub-standards up to 100,000 ohms, and below 1 ohm to $\frac{1}{1000}$ ohm.

The expression " meter" includes a large number of instruments which formerly were only regarded as scientific measuring instruments, but which now have come into common use, as volt-meters, am-meters, &c.

The Weights and Measures Act, 1889 (s. **6**), provides that the Board of Trade shall from time to time cause such new denominations for the measurement of electricity, temperature, pressure, or gravities, as appear to them to be required for use in trade to be made and duly verified; and in 1891 and 1892 a Committee on Standards for the **Measurement of Electricity, appointed by** the Board of **Trade, made important recommendations** (Reports, 1891 **and 1892) to which effect was given by an Order in Council dated 23rd August 1894,** the standards of the ohm, ampère, **and volt, being legally** defined **for use in trade.** To the Order is appended a specification dealing with the " Clark Cell."

Recommendations in Report of Electrical Standards Committee, 1892.

1. That it is desirable that new denominations of standards for the measurement of electricity should be made and approved by Her Majesty in Council as Board of Trade standards.

2. That the magnitudes of these standards should be determined on the electro-magnetic system of measurement with reference to the centimetre as unit of length, the gramme as unit of mass, and the second as unit of time, and that by the terms centimetre and gramme are meant the standards of those denominations deposited with the Board of Trade.

3. That the standard of electrical resistance should be denominated the ohm, and should have the value 1,000,000,000 in terms of the centimetre and second.

4. That the resistance offered to an unvarying electric current by a column of mercury at the temperature of melting ice 14·4521 grammes in mass of a constant cross sectional area, and of a length of 106·3 centimetres may be adopted as one ohm.

5. That a material standard, constructed in solid metal, should be adopted as the standard ohm, and should from time to time be verified by comparison with a column of mercury of known dimensions.

6. That for the purpose of replacing the standard, if lost, destroyed, or damaged, and for ordinary use, a limited number of copies should be constructed, which should be periodically compared with the standard ohm.

7. That resistances constructed in solid metal should be adopted as Board of Trade standards for multiples and submultiples of the ohm.

8. That the value of the standard of resistance constructed by a committee of the British Association for the advancement of science in the years 1863 and 1864, and known as the British Association unit, may be taken as 0·9866 of the ohm.

9. That the standard of electrical current should be denominated the ampère, and should have the value of one-tenth (0·1) in terms of the centimetre, gramme, and second.

10. That an unvarying current which, when passed through a solution of nitrate of silver in water, in accordance with the specification attached to this Report, deposits silver at the rate of 0·001118 of a gramme per second, may be taken as a current of one ampère.

11. That an alternating current of one ampère shall mean a current such that the square root of the time average of the square of its strength at each instant in ampères is unity.

12. That instruments constructed on the principle of the balance, in which by the proper disposition of the conductors, forces of attraction and repulsion are produced, which depend upon the amount of current passing, and are balanced by known weights, should be adopted as the Board of Trade standards for the measurement of current whether unvarying or alternating.

13. That the standard of electrical pressure should be denominated the volt, being the pressure which, if steadily applied to a conductor whose resistance is one ohm, will produce a current of one ampère.

14. That the electrical pressure at a temperature of 15° Centigrade between the poles or electrodes of the voltaic cell known as Clark's cell, prepared in accordance with the specification attached to this report, may be taken as not differing from a pressure of 1·434 volts, by more than one part in one thousand.

15. That an alternating pressure of one volt shall mean a pressure such that the square root of the time average of the square of its value at each instant in volts is unity.

16. That instruments constructed on the principle of Lord Kelvin's Quadrant Electrometer used idiostatically, and, for high pressures, instruments on the principle of the balance, electrostatic forces being balanced against a known weight, should be adopted as Board of Trade standards for the measurement of pressure, whether unvarying or alternating.

In connection with measuring instruments thermometers are used, and in the measurement of temperature for weights and measures purposes, the Fahrenheit scale (32° to 212° F.) is legally adopted; for instance, the Imperial yard has its true length at 62° F.; the gallon measure contains its true weight of water at 62° F.; and the unit of measure in the sale of gas is the cubic foot at 62° F., &c. All metric weights and measures are, however, referred to the Centigrade scale (0° to 100° C.) and the Centigrade scale is also used in connection with electrical standards.

Measurement of temperature.

There is no legal standard of the Fahrenheit thermometer, but three representative thermometers, originally
verified by Sheepshanks, and one presented by the Kew
Committee of the Royal Society in 1869 ; and also a standard
centigrade thermometer; are deposited at the Standards
Office.   The relation of the Fahrenheit thermometer to the
Centigrade scale in terms of the hydrogen thermometer
has been thereby so far determined by the Comité
International des Poids et Mesures, 62° F. being taken as
equivalent to $16\overset{\circ}{\cdot}667$ C. The Centigrade scale, has had
its degree carefully defined : a " normal degree " is a degree
of the Centigrade thermometer when the point 100° C.
corresponds to the temperature of pure boiling water under
the pressure of a column of mercury ($\Delta = 13\cdot59593$) at
the height of 760 millimetres at 0° C., in latitude 45° and
at the level of the sea ; the zero point of the scale being
at the temperature of melting ice.

One of the first attempts made in this country to
establish accurate Fahrenheit thermometers was probably
that made by the Rev. R. Sheepshanks, F.R.S., in 1847, before
which time it would appear that there was no workman in
London who could be trusted to make a good thermometer.
Sheepshanks' standards were followed by those of Balfour
Stewart (Kew), and of Clarke (Comparisons of Standards,
1866).

The old form of mercurial thermometer, having spherical
bulbs (like Sheepshanks'), or containing a large mercurial
column, has been now superseded for standards purposes
by the new form of thermometer adopted by the Comité.
The above standard Centigrade thermometer has a
cylindrical bulb ($1\cdot7 \times 0\cdot2$ inch) with a very fine column of
mercury; it was made in 1890, and was originally verified
by the Comité in relation to the hydrogen thermometer.
Besides this standard there are two other similar standards
which accompany the metres prototypes. This thermometer is about 702 millimetres long ; the length of a normal

degree being 5·3 mm.; and it is divided throughout into
0°·1 C. from 4° to 103°. Its error at the boiling point was
0°·02 C. and **at** the zero point 0°·07 C. when placed in a
horizontal position, and there is also a small correction for
external atmospheric pressure; and for reduction to sea
level, and to the normal latitude of 45° (*vide* Guillaume's
Thermométrie de Précision, p. 30). When placed in a
vertical position the "co-efficient of **pressure**," or correction
for the internal pressure of the mercury for one millimetre
of mercury is only 0·0001. This **thermometer** is **made**
of hard glass **(Tonnelot) in which** the **zero point** is
practically constant, **showing** on analysis 71·5 **silica, 14·5**
lime, **11·2 soda, 1·3 alumina, 0·7** sulphuric acid, 0·3 **potash,**
and per-oxide of **iron, 0·3.**

Mr. Sheepshanks **has** given an account in the Monthly
Notices of the Royal Astronomical Society for June 1852
(reprinted by George Barclay, Leicester Square, 1852), **of**
his proposed method of constructing standard thermometers.
He thought that he could produce instruments which, with
certain precautions, would show a thirtieth **or even a**
fiftieth of a degree Fahrenheit, but he found that, after
determining the boiling point, the freezing point of a
thermometer would fall from one-third **to** one-fourth of a
degree.

The International Meteorological Tables (Paris, 1890), Barometer
published in conformity with the resolution of a Congress scales.
at Rome (1879), state, with reference to the comparison
of the different barometrical scales, as follows (Chap. **4,**
p. B. 23):—

"The correct reading of the height of the barometer supposes a two **fold**
correction **of** temperature to bring on the one hand the divisions of the
scale to the normal temperature of the standard of length, and on the other,
the mercury in the barometer **to a fixed temperature, viz., the melting point**
of ice.

Thus in France, the mercury and the scale are set to 0° C., whilst in England
the mercury is set to 32° F., and the scale to 62° F. which is the normal

temperature of the standard of length. Under these conditions the coefficient of conversion for barometrical measurements is the same as for linear."

As the metre has its true or legal length at 0° C., and the yard has its true or legal length at 62° F., we can only find the true or legal difference between the two measures when we take the one at 0° C., and the other at 62° F.

Hydrometers. For the purpose of determining the degree of gravity of spirits the only hydrometer known to the law is that of "Sikes," and Mr. R. Bannister, the Deputy Principal of the Government Laboratory at Somerset House has kindly furnished the writer with the following information with reference to Sikes' hydrometer.

The hydrometer at present used for levying the duty on spirits is that of Sikes, which was legalised by Act 58 Geo. 3 c. 28 (1817). It is made of brass, gilt, the spherical bulb being fitted with a stem at either pole. The upper stem bears the scale of 10 main divisions, each of which is again divided into five parts, giving 50 sub-divisions in all. The lower stem serves to secure upright flotation and to carry the weights by means of which the range is extended to 500 sub-divisions representing about 185 degrees of gravity. Sikes' instrument does not, however, indicate specific gravity (sp. gr.), but shows by the help of tables (Tables for ascertaining the Strength of Spirits with Sikes' Hydrometer) the "strength" over or under "proof" of the sample examined. A definition of "proof" is given in terms of sp. gr. in the Act as that which at 51° F. weighs exactly twelve-thirteenths of an equal bulk of distilled water. The technical terms "over" and "under" proof will be readily understood from the following :—

Spirit is $x°$ overproof (o. p.) when 100 gallons of it will produce $(100 + x)$ gallons at proof by dilution with water.

Spirit is $x°$ underproof (u. p.) when 100 gallons of it contain $(100 - x)$ gallons of proof spirit.

For raising the duty on beer a direct sp. gr. instrument, or gravimeter with poise weights, termed Bate's Saccharometer, is in use. It is a gilt brass instrument, constructed on the same principle as the hydrometer, and shows specific gravities at 60° Fahrenheit. By means of an admirable contrivance in the size and weight of the poises the same stem is used for 180 degrees of specific gravity. This instrument and tables for temperature corrections have been in use since 1829.

Turning now to the means adopted for verification, the Saccharometer, showing as it does the sp. gr. of the solution in which it floats, is controlled directly by the gravity bottle; the hydrometer is compared with the standard instrument, which in turn is controlled by comparison of its weight and volume with data derived from Sikes' original instrument.

The alcoholmeter used by Her Majesty's Customs is also Sikes' hydrometer, as the Customs laws impose a duty on spirits according to the proof gallon, and it is necessary to use an instrument formed so as to show the gravity of any mixture of alcohol and water in which it is immersed.

The brass bulb of a hydrometer, being thin and hollow, renders the instrument liable to injury from abrasion or rough treatment, and from the incrustation of the colouring matters of the liquids tested. From any of these causes it is easily put out of order, so that its accuracy requires to be carefully tested from time to time by comparison with standard instruments. When a Sikes' hydrometer is inaccurate to the extent of a quarter of one degree marked on the stem, it is considered to be out of order.

The United States Government adopted standard hydrometers made of glass and not of metal, but glass instruments are of course more easily broken, although the ordinary hydrometers and saccharometers, &c., used by chemists are mostly made of glass.

Other hydrometers are also used in determining the specific gravities of various liquids ; for instance, Beaumé's hydrometers (1) for fluids heavier than water, (2) for fluids lighter than water, and M. Francœur has given the following table of specific gravities corresponding with Beaumé's hydrometer at $54\frac{1}{2}°$ F.

(1)		(2)	
Beaumé degree.	Specific gravity (pèse-acide).	Beaumé degree.	Specific gravity (pèse-esprit).
0	1·0000	10	1·0000
5	1·0340	15	0·9669
10	1·0704	20	0·9359
15	1·1095	25	0·9068
20	1·1515	30	0·8795
25	1·1968	35	0·8538
30	1·2459	40	0·8295
35	1·2992	45	0·8066
40	1·3571	50	0·7849
45	1·4206	55	0·7643
50	1·4902	60	0·7449
55	1·5671		
60	1·6522		
65	1·7471		
70	1·8537		
75	1·9740		

The centesimal alcoholmeter of Gay Lussac is also used ("Instruction pour l'usage de l'Alcoomètre Centésimale." Par M. Gay Lussac. Paris, 1824). To determine the quantity of water contained in ardent spirits, Gay Lussac took pure alcohol by volume at 15° C. and represented its strength by 100 per cent., or unity ; hence the strength of a liquor is simply the per cent. by volume of pure alcohol which it contains at 15° C., at which temperature this hydrometer is graduated from 100 (alcohol) to 0 (water), each graduation or degree of the scale representing a per cent. of alcohol.

Tralles' centesimal alcoholmeter differs very slightly from Gay Lussac's. The specific gravity of water is taken by Tralles as unity at its maximum density (4° C.), so that it becomes 0·9994 at 60° F.; whereas Gay Lussac assumed the unity of water at 15° C.; and took the specific gravity of absolute alcohol at 0·7947 (t = 15° C.) while Tralles took 0·7939 (t = 60° F.). Tralles based his result (1811) on the weighings of water by Gilpin and Blagden (Phil. Trans. Roy. Soc. 1790) and Gay Lussac on his own experiments in 1824. **Both** weighings of water have, however, been now superseded (*see* p. 88) by more accurate weighings (*see also* " Mémoire sur l'aréometrie." **Francœur, Paris, 1842.** "United States Reports on **Sugar and Hydrometers**" made by Prof. A. D. **Bache** and **Prof. R. S. McCulloh.** Washington, 1848. **Ure's** "Dictionary of Arts and Manufactures," 1843, p. 23. Art. "alcohol ").

Indirectly derived from scientific measuring **instru**ments are the standards used in the sale of gas, regulated by the Sale of Gas Act, 22 & 23 Vict. **c. 66, 1859**; 23 & 24 Vict. **c. 146, 1860**; and the Gas **Works Clauses Act** (1847) **Amendment, 1871.** The Act of 1859 provides that the only legal **standard or unit of measure for the** sale of gas by meter **shall be the** cubic foot containing 62·321 lbs. avoirdupois **weight** of distilled or rain water weighed in air at the temperature of 62° F. thermometer, the barometer being at 30 inches.

Gas-measuring Standards.

The actual Board of Trade standards for the measurement of gas are known as "gasholders," and in 1861 under the direction of H.M.'s Treasury three such standard gasholders of the cubic foot, 5 cubic feet and 10 cubic feet, were deposited with the Comptroller-General of the Exchequer, and were subsequently transferred to the Board of Trade in 1866. Copies of these models were also sent in 1862 to the Lord Mayor of London, and to the chief magistrates of Edinburgh and Dublin respectively; and other copies have since been obtained by the local authorities of other cities and boroughs. A standard of 20-cubic feet was also added in 1893.

By means of such copies the local inspectors of gas meters, who are appointed by the local authorities, test the gas meters fixed on consumers' premises. No practical alteration has taken place in the mode of measuring gas since these standards were made in 1861; excepting in some improvements in detail, the same forms of "wet" meter and "dry" meter being still used in the sale of gas. On every consumer's meter tested and stamped by an inspector a stamp of the following official design should appear.

Fig. 15.

Besides the official stamp of verification the stamp of the local authority is placed by the inspector on the meter. The Sale of Gas Act, 1859, allows an inspector to pass any meter which is not more than 3 per cent. in favour of the consumer and 2 per cent. in favour of the seller of gas, so that two stamped meters might vary 5 per cent.

The standard cubic foot itself is actually determined by means of a conical copper vessel or "bottle" holding a cubic foot of water, and not by means of any vessel of rectangular shape; the probable error of measurement being less in the use of a measure of conical shape holding a cubic foot of water than in a cubed measure of the dimensions of 1 foot.

Fig. 16.—GAS-ROOM, STANDARDS OFFICE.

Showing the 10, 5, and 1 cubic foot Standards.

In Fig. 16 the three gasholders, 10, 5, 1 cubic feet, are shown on **the right of the picture as** cylindrical measures suspended **over tanks of** water.

Two cubic foot **" bottles " are** shown **on** the left of the picture, one on either side of the clock ; each conical bottle being also suspended over a water tank.

Every gas meter inspector's testing instruments are required to be reverified **once in every 10** years (Weights and Measures Act, 1889, section 15).

In the measurement **of the illuminating** power **of gas** Photometers. —rather a qualitative **than a** quantitative measurement— the Gas **Works** Clauses **Act, 34 & 35** Vict. **c. 41, 1871,** prescribes **the present form of photometer to be used by** gas companies in testing **the power of** gas as **the** " improved form **of** Bunsen's **photometer "** known either **as " Letheby's "** open **60 inch** photometer, or as **" Evans' "** enclosed 100 inch photometer, as provided with a proper meter, minute clock, governor, pressure gauge, and balance.

There is no standard photometer, but **such** instruments as the meter, **balance, and pressure - gauge, &c. are** standardized. (*See also* Dib lin's Photometry, "Journal of Gas Lighting," **London 1889;** Hartley's **" Gas** Analyst's Manual," Spon, London **1879, &c.).**

A useful form **of photometer is that** introduced by **Mr.** G. Lowe, **C.E., in 1864, which** simply indicates the illuminating power of **gas from** the quantity required under specified conditions to maintain a jet of gas or **a** flame a given **height—as 3** inches.

There **is no standard** of a " Burner " except in some cases of **" Sugg's London** Argand Burner," specified in local Acts, and where the Board of Trade are required to certify a model of **such** burner, then the stamp of the Standards Office is placed thereon.

In connection with **the** Argand burner **Mr. J**ohn Methven, C.E., has introduced a form of " screen " by means of which certain uniform photometric results can be obtained in the use of a burner.

As to a standard of light, it may be pointed out that, so far back as the year 1849, a standard candle was recommended by Professor Grabam, Dr. Leeson, and Mr. Cooper. The Metropolis Gas Act of 1860 further provided that the common gas supplied shall be "such as " to produce from an argand burner having 15 holes and " a 7 inch chimney, consuming 5 cubic feet of gas an " hour, a light equal in intensity to the light produced by " not less than 12 sperm candles of six to the pound, each " burning 120 grains an hour."

The Gas Works Clauses Act, 1871, sections 28 and 32, prescribes the mode of testing the illuminating power of gas supplied by gas companies, and in London the Metropolis Gas Referees issue Regulations with reference thereto.

In a Report made by a Committee on the Standards of Light in 1895 and laid before Parliament, a new form of standard of light known as the " pentane " standard burner, has been proposed in place of the sperm candle.

Weighing of air.   The Imperial Pound, the unit of mass, has its true legal weight " in vacuô " and hence the weight of air displaced by it, and by any mass compared with it, has to be allowed for. It has been remarked by Dr. Angus Smith (" Air and Rain ") that when we are children, air is to us, nothing— a vessel of air is a vessel with nothing in it—and that early nations thought in the same way. It was a great day for the world when air was found to be something material, and to be capable of weighing down the scale of a balance.

The weight in grains of a cubic foot of air, when the atmospheric pressure is B, expressed in inches, and the temperature t, in degrees **Fahrenheit**, may be taken as follows :—

$t.$ Fahrenheit.	B. Inches.	Weight in Grains of a Cubic Foot of Air.		
		Dry Air.	Ordinary Air (Saturation $=\frac{2}{3}$).	Moist Air. (Saturation $= 1$).
		Gr.	Gr.	Gr.
32	30	566·890	566.491	565·587
62	30	534·214	531·720	530·474
80	30	516·376	511·921	509·667

In the measurement of any given **volume** of air we have to allow, **not** only for variations of atmospheric pressure (**B.**) and of temperature ($t.$), but also for the variable amount of moisture contained in the air, and for the amount of carbonic acid. The weights above given refer to the air of rooms artificially warmed and illuminated, which contain **four** volumes of **carbonic acid in every** 10,000 volumes of air. **Some small corrections** are necessary for the accelerative **effect** of the **force of** gravity (g) at the latitude of **the place where the air is** weighed (the latitude of **45°, being referred to as a normal** latitude for such weighings),

$$\left.\begin{array}{c} g. \\ \text{London} \end{array}\right\} = \begin{array}{c} g.\ 1\cdot000577. \\ 45° \end{array}$$

and also for the height of the place above sea level. **The** above weights of the cubic foot are those for the latitude of Westminster, **51° 29′ 53″, at** 5 metres above mean sea level (as determined by the mean level **at** Liverpool). In such latitude **a** *litre* **of** dry air would also weigh 1·293934 grams ($t = 0°C.$, B $= 760$ millimetres), but **a** litre of such air **freed from carbonic** acid would **weigh** 1·293519 grams.

G 2

With reference to the use of the expression "mass" it would appear that the expression "weight" is generally used in legal enactments.

Weighing
of water.

The weight of a cubic foot of water has been fixed by the present Sale of Gas Act, 1859, but the Weights and Measures Acts do not fix the actual weight of a given volume of water, as of a cubic foot. The weight adopted in the Sale of Gas Act was derived from the evaluation of the mass of a cubic inch of distilled water based on weighings made in 1798 by Sir G. Shuckburgh (Philosophical Transactions, Royal Society, 1798, p. 133); and on measurements made in 1821 by Captain Kater (Phil. Trans., 1821, pp. 316 and 326). According to these measurements a cubic inch of distilled water weighed 252·458 grains, and the capacity of the Imperial Gallon was taken at 277·27384 cubic inches.

The Imperial Standard Bushel, which is equal to 8 gallons, had, therefore, an apparent capacity of 2218·19072 cubic inches. As the diameter of the standard bushel is double its internal depth, its diameter would be 17·80927948 and its depth 8·90463974 inches.

A more recent evaluation of the mass of a cubic inch of distilled water has been stated in the Transactions of the Royal Society, pp. 331–354, 1892, showing that the mass of the cubic inch of distilled water, *freed from air*, weighed in air at the temperature of 62° F., the barometer being reduced to 30 inches, against brass weights of the density of 8·143, was found to equal 252·286 grains ± 0·0002 grain. A cubic foot of such water under similar conditions would weigh 62·278601 lbs. avoirdupois. It appeared that a cubic foot of distilled water freed from air weighed about 321 grains more than when "saturated" with air. The value 252·286 is adopted in an Order in Council dated 28th November 1889, legalising certain new measures of capacity.

Where distilled water is not available, or pure rain water, then ordinary water may be used :

	Grains.
One gallon once distilled water weighs -	70000·5
„　　twice　„　　„　　„　-	70000·0
„　　well water weighs　-　-	70066·6

Professor D. Mendeléeff has also given the following as the weight of a cubic decimetre of distilled water freed from air :—

Temperature on the Hydrogen Thermometer Scale.	Weight of Water in Vacuo		
	Of a cubic decimetre in grains.	Of a cubic inch in grains.	Of a cubic inch in Russian dolias.
4° max. den. distilled water. 16⅔° C. equivalent to 62° F.	999·847 998·715	252·854 252·568	368·734 368·316

(Proceedings, Royal Society, London, February, 1896, p. 155.)

In ancient times, although some attention was given to the purity of the water used when determining the capacity of a standard measure of capacity, Dr. Wybard was probably the first English experimenter who took especial care in the selection of pure water (Wybard's "Tactometria," London, 1650), as in his experiments in London on the capacities of the standard ale and wine gallons of 1650, he used rain water "as it fell from the clouds," and snow water, as well as the waters of the Thames and the "new river of Ware," and the springs of Lamb's Conduit, Tower Hill, and Cripplegate.

Com-
parisons of
standard
linear
measures.

In the verification of the standard of length, or in the comparison of graduated or linear standards of length, the form of comparator generally used is one fitted with two micrometer microscopes, which are placed over the two graduations or lines between which any given distance or interval of length is to be measured. Such microscopes are firmly fixed so that when an interval of length has been measured on one comparing standard the other standard measure can be brought under the microscopes and the corresponding interval of length marked on it can be thus compared with that on the comparing standard (*vide* Fifth Report, Standards Commission, Appendix vii.) The two micrometer microscopes are thus used much in the same way as a pair of beam compasses might be, with the important exception that the standards are not touched or in any way interfered with whilst under the microscopes : both standards being placed in a box so that a uniform temperature may be secured during the comparison. In the following figure (Fig. 17) the two vertical micrometer microscopes are shown as mounted in position and fixed to a stone shelf which is supported by a platform of masonry standing on a concrete base. Under the microscopes is a travelling platform on which rests the long iron box or trough shown in the middle if the figure (Fig. 17).

The following will show the amount of accuracy within which standard yards can be constructed. In 1882 the three Parliamentary copies of the Imperial Yard were compared one with the other as follows :—

    Yard No. 2   Yard No. 3 − 0·000091 inch.

    Yard No. 2 = Yard No. 5 − 0·000029 inch.

    Yard No. 3 = Yard No. 5 + 0·000062 inch.

the difference in no case reaching one ten-thousandth of an inch. For the above comparison a large number of

FIG. 17.—COMPARATOR FOR STANDARDS OF LENGTH, &c., &c.

observations were made, and the results of these comparisons were calculated as follows :—

ACTUAL DIFFERENCES in parts of an Inch between **the three** STANDARD YARDS Nos. 2, 3, and 5.

—	—	No. 2 = No. 3.	No. 2 = No. 5.	No. 3 = No. 5.
		Inch.	Inch.	Inch.
(a)	Arithmetical mean of the observations -	−0·000089	−0·000032	+0·000066
(b)	Results obtained by allowing for the value of each comparison according to its probable error, but without combining the three final results -	−0·000093	−0·000028	+0·000061
(c)	Results obtained by combining the three results and allowing for their probable errors -	−0·000091	−0·000029	+0·000062

The **results** (c) finally adopted being obtained by combining the three separate **results** (a) and allowing for their probable **error** ($\pm$ 0·000012 inch) as follows, the three **sets of** comparisons being made independently of each other, so that **any error** committed in one set did not affect the **other set** :

$$A_1 = A_2 + a_1 - \frac{a_1 + a_2 + a_3}{e_1^2 + e_2^2 + e_3^2} e_1^2$$

$$A_2 = A_3 + a_2 - \frac{a_1 + a_2 + a_3}{e_1^2 + e_2^2 + e_3^2} e_2^2$$

$$A_3 = A_1 + a_3 - \frac{a_1 + a_2 + a_3}{e_1^2 + e_2^2 + e_3^2} e_3^2$$

Here A_1, A_2, A_3, represent **the three yard measures**; a_1, a_2, a_3 their differences : e_1, e_2, e_3 **their probable errors.**

All comparisons are referred to a temperature of 62° F. for the yard, and 0° C. for the metre, therefore the actual temperature at the time of observation has to be taken, and a correction made if the temperature is not exactly 62° F., or 0° C.

Examples of co-efficients of linear expansion as applied to the corrections for temperature :—

		For 1° F.	For 1° C.	—
Brass, Cast - - -		0·00000957	0·00001722	Sheepshanks.
Bronze, Baily's - -				
Copper 17 - -		0·00000986	0·00001774	Clarke.
Tin 2½ - -				
Zinc 1 - -				
Copper - - -		0·00000887	0·00001596	Fizeau.
Glass (Thermometer) -		0·00000499	0·00000897	Regnault.
Gold - - -		0·00000786	0·00001415	Roberts-Austen.
Iron, Wrought - -		0·00000648	0·00001166	Clarke.
Iron, Cast - -		0·00000556	0·00001001	Sheepshanks.
Platinum - -		0·00000479	0·00000863	Fizeau.
Iridio platinum—				
90 °/₀ platinum -		0·00000476	0·00000857	,,
10 °/₀ iridium -				
Silver, Pure - -		0·00001079	0·00001943	Roberts-Austen.
Slate - - -		0·00000577	0·00001038	Adie.
Steel, Cast - -		0·00000636	0·00001144	Clarke.
Steel, Tempered -		0·00000689	0·00001240	Fizeau.
Stone, Dry (Sandstone) -		0·00000652	0·00001174	Adie.
Tin - -		0·00001163	0·00002094	Fizeau.
Wood (Pine) - -		0·00000276	0·00000496	Kater.

In the work of dividing or graduating a linear standard a "Dividing-Machine" is employed, and the graduations can only be accurately made when a screw of great precision is used, so set as to give equal divisions of the scale. The graduation of the Imperial standard was carried out by a machine designed by Messrs. Troughton and Simms. In the production of "diffraction gratings" used in optical research as in the measurements of wave-lengths of light, the highest precision has been obtained in the use of the dividing machine; rulings of 20,000 lines to the inch being

FIG. 18.—BALANCE ROOM, JEWEL TOWER.

cut by a diamond on speculum metal, only visible of course with a microscope of high power. The unit of a wavelength is the ten-millionth of a millimetre.

For scientific work balances of great precision are used, and the use and theory of such balances has been well explained by several authors, as— *Balances of precision.*

HARTIG, Dr.—" Theorie der gleicharmigen Waagen, Polyt : Centralblatt," 1859.

JEVONS, Prof. — Article " Balance " in Watt's " Dictionary of Chemistry."

DITTMAR, Prof. Wm., F.R.S.—Article " Balance " in 9th edition of the " Encyclopædia Britannica."

MAREK, W. J.—" Formules, Constantes et Tables, Travaux et Mémoires du Bureau International des Poids et Mesures," Tome III.

THIESEN, Dr. M.—" Études sur la Balance, Travaux et Mémoires du Bureau International des Poids et Mesures," Tome V.

WALKER, JAMES, M.A.—" The Theory and Use of a Physical Balance." Oxford, 1887.

STEWART, BALFOUR, F.R.S., and GEE, W. W. H.— " Lessons in Elementary Practical Physics." London, 1889. Vol. I., Chapter III.

The balances of precision used at the Standards Office are kept in the basement of the Jewel Tower (*see* page 123), and are placed on slate shelves or on pedestal tables, as shown in Figure 18 ; they will " turn " with—

No. 1.	300 lbs. avoir.	in each pan will turn with			1	grain
No. 2.	50 lbs.	„	„	.,	0·2	„
No. 3.	10 lbs.	,.	„	,.	0·05	„
No. 4.	5 lbs.	,.	,.	,,	0·005	,.
No. 5.	1 lb.	,,	,,	,.	0·002	„
No. 6	500 grains	,	.,	„	0·0005	„
No. 7.	10 „	,,	„	„	0·0002	„

In accurate weighings it is desirable to be able to change the weights from one side of the beam, or the pans and weights together, without opening the balance case or disturbing the beam, and this can be done by a simple method introduced by Professor W. H. Miller, who, by means of forked rods inserted through small holes in either end of the balance-case, was able to lift each pan with its weight off its bearing on to a carrier running at the back of the beam, and so to carry or transfer the pans to either end of the beam.

Another and better method is that adopted in the "Ruprecht" balance, used by the International Committee of Weights and Measures at Paris in the verification of kilogram weights (*see* "Travaux et Mémoirs du Bureau International." Tome 1, p. D. 53, 1881).

Balances for weighing in vacuo, so as to avoid correction for the weight of air displaced by bodies of various densities, are not much used, as there is less probability of error in calculating the weight of air displaced by a body under given conditions as to temperature, &c. than in making actual use of the vacuum balance. There are, however, two well-known forms of vacuum balances, Miller's large balance, and a smaller and more convenient balance for chemical work known as Mendeléeff's, the distance between the knife edges of which is only $4\frac{1}{2}$ inches, and it is possible with it to weigh 500 grammes.

The balance used in 1892 at the Houses of Parliament in the re-verification of the Imperial Pound, showed differences of weight equal to 0·0002 grain. This balance is enclosed in a copper case of the form shown in Fig. 19, as suggested by Professor C. V. Boys, F.R.S. having the following dimensions :—

Height of case   = $18\frac{7}{8}$ inches.
Width   ,,   = $21\frac{3}{4}$,,
Depth   ,,   = $4\frac{7}{8}$,,

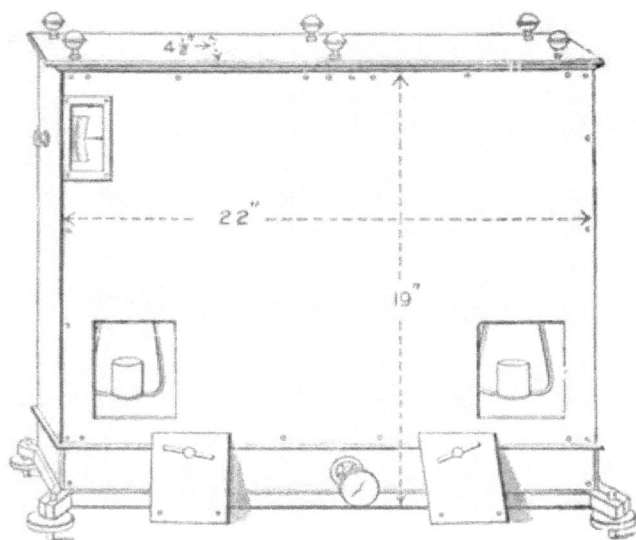

Fig 19.—BALANCE OF PRECISION ENCLOSED IN A COPPER CASE.

PART V.

14. THE METRIC SYSTEM.

Description of new Standard Metre and Kilogram.

Figures 20 and 21 show the forms as hereafter described of the new metric standards of length and mass respectively (prototypes nationaux), delivered to the Board of Trade by the International Committee of Weights and Measures at Paris on 28th September 1889, a third and final standard being received from the Committee in December 1894.

The two standards received in 1889 include a "line" standard metre measure (mètre-à-traits) and a kilogram weight. The standard received in 1894 is an "end" standard metre (mètre-à-bouts). These three standards, together with other similar standards supplied to 21 different States, are (inter alia) the outcome of the results of the labours of the International Committee for nearly 20 years.

The standards were verified at the Bureau International des Poids et Mesures (Pavillon de Breteuil, Sévres, près Paris), which bureau was established under a Metric Convention, dated 20th May 1875, signed by 20 different High Contracting States, exclusive of Great Britain, who finally joined the Convention in September 1884.

The Committee is founded and maintained by common contribution from all countries who are parties to the Convention. The Committee was charged in 1875 with the construction, restoration, and verification of new metric standards (prototypes internationaux) to replace the original metric standards of France (mètre et kilogramme des archives), and with the verification of copies of the new standards for all the contracting States. By such means the international accuracy of metric standards is now assured throughout the world.

Iridio-platinnm metre.

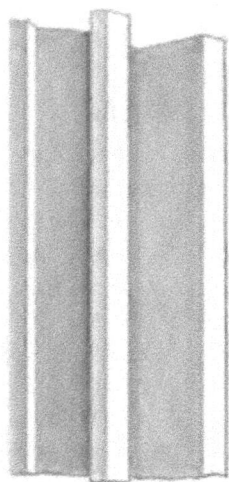

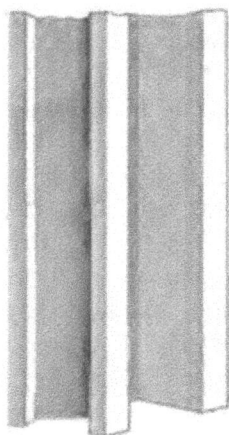

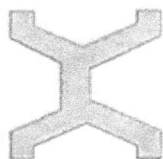

Fig. 20.—NEW STANDARD METRE.

b.

Fig. 21. NEW STANDARD KILOGRAM.

The Mètre.

The two metric standards above referred to are made of iridio-platinum, or an alloy of 90 per cent. of platinum and 10 per cent. of iridium. The metres are in transverse section, nearly of the form of the letter X, known as the Tresca form (*see* section *a* and length *b*, Fig. 20) an economic form (iridio platinum being a costly metal), and one less affected by heat, practically non-oxidizable, and well adapted for receiving finely engraved lines. In fact the alloy appears to be of all substances the least likely to be affected by time or circumstance, and has been preferred for standards purposes to rock crystal, gold, &c. The lines on the mètre-à-traits are very fine, and are barely visible to the naked eye.

The iridio-platinum standards were constructed in London under the care of George Matthey, Esq., F.R.S., and were finished by Messrs. Brunner, M. Collot, and M. Laurent, of Paris. The lines on the metre were traced by M. G. Tresca, at the Conservatoire des Arts et Mètiers, Paris. The chemical tests of the alloys were made by Professor Stas, and by MM. H. St.-Claire-Deville, Debray, and Tornöe. The actual verifications were partly made by Dr. Broch, but mainly by Dr. Benôit, with the assistance of M. Marek, Dr. Pernet, Dr. Chappuis, Dr. Guillaume, and M. Thiesen.

The actual relation of our prototype metre No. 16 is as follows :—

At 0° Centigrade :

No. $16 = 1$ metre $- \overset{\mu}{0\cdot6} \pm \overset{\mu}{0\cdot2}$ at $0°$ C.

Here μ means one micron, or one-thousandth of a millimetre (or nearly $0\cdot00004$ inch), so that metre 16 may be said to have been verified with an accuracy of one part in five millions.

The "end" standard bar, or mètre-à-bout has been verified also with great accuracy, with a probable error of $\pm\overset{\mu}{0\cdot3}$. In the verification of the end standard, MM. Cornu and Benôit originally introduced a method of reflection, by means of which it was unnecessary to bring the ends of a bar into contact with any touching surfaces, and thus the measuring ends of a bar might be carefully preserved and used.

The Kilogram.

The unit of mass of the kilogram (Fig. 21) is determined by a solid piece of iridio-platinum in the form of a cylinder, the height and diameter of which are equal (39 millimetres). The kilogram, No. 18, supplied to Great Britain has a faint distinguishing mark, and is highly polished. On analysis, the metal showed very faint traces of ruthenium, rhodium, and iron. Its volume (18) was found to be at 0° Centigrade—

millilitres.

$18 = 46\cdot414.$

corresponding to a density of—

$21\cdot5454.$

After its final adjustment it was found to be in vacuô at 0° Centigrade.

kg. milligram.

Prototype $18 = 1 + 0\cdot070 \pm 0\cdot002.$

WEIGHINGS REQUIRED. (1) to (6).

Example.

Weighings										Mgr.
(1)	500	200	200'	100	=	Kilo	−	44·458		
(2)		200	200'	100	=	500	+	49·940		
(3)		200			=	200'	−	0·250		
(4)	50	20	20'	10	=	200	+	1·606		
(5)	100	50	20	20'	10	=	200'	+	1·479	
(6)	100	50	20	20'	10	=	100	+	1·037	

Reductions.

200'	=	200	+	0·250		
10	=	200'	+	1·606		
10	=	200'	+	1·479		
2 (100)	=	2 (50, &c.)	−	2·074		
4 (100)	=	2 (200)	+	1·261		
2 (100)	=	200	+	0·6465		
2 (100)	=	200'	+	0·3805		
100	=	100	+	0·0000		
5 (100)	=	200 200' 100	+	1·011		
		200 200' 100	+	2·741		
∴ 5 (100)			=	+3·752		
Therefore 100			=	+0·7504 mgr.		

Series.

500
200
200' }
100

Mgr. :

500	200	200'	100	=	500	=	44·458
	200	200'	100	=	500	=	+ 49·940
2 (	200	200'	100)	=		=	+ 5·482
	200	200'	100	=		=	+ 2·741
			& 500	=		=	− 47·199

The above process being repeated for each set, 50...10
5...1
0.5...0.1 and so on.

So that it may be said that the kilogram (kg.) has been verified with a probable accuracy of 0·002 parts in a million.

Besides the primary metric standards, we have also iridio-platinum standard weights (1895) below the kilogram, as follows :—

Standard Weights below the Kilogram.

500 grams.
200 „
200' „
100 „
50 „
20 „
20' „
10 „

down to 0·001 gram or a milligram.

In order to obtain the true weight of each weight in the series 500, 200, 200', 100, &c. it is of course necessary to compare all four weights with the kilogram prototype (1,000 grams); and then the 50, 20, 20', and 10 with the the 100 grams, and so on down to the 0·001 gram. For such purpose a separate series of weighings is required; and the six weighings (1) to (6) when obtained are reduced according to the formula given opposite (p. 100).

The equivalents of the metric trade weights and measures, in terms of the Imperial system, are (Weights and Measures Act, 1878, sections 18 and 21) as follows :—

IMPERIAL TO METRIC (1878).

Metric Equiva-lents, 1878.

LINEAR MEASURE :
1 Yard (3 ft.) - - = 0·91438348 Metre.
1 Mile (1,760 yds.) - - = 1·60931493 Kilometres.

H

SQUARE MEASURE :
  1 Square Yard (9 sq. ft.)   - = 0·83609715 Sq. Metre.
  1 Acre (4,840 sq. yds.)   - = 0·40467 Hectare.

CUBIC MEASURE :
  1 Cubic Inch   -   - = 16·38617589 Cub. Centimetres.
  1 Cubic Foot (1,728 cub. in.) = $\begin{cases} 0\cdot02832 \text{ Cub. Metre, or} \\ 28\cdot31531 \text{ Cub. Decimetres.} \end{cases}$

MEASURE OF CAPACITY :
  1 Gallon (4 qts.)   -   - = 4·54845797 Litres.

AVOIRDUPOIS WEIGHT :
  1 Pound (16 oz. or 7,000 grs.) = 0·45359265 Kilogram.

TROY WEIGHT :
  1 Troy Ounce (480 grs. avoir.) = 31·10350 Grams.
  1 Pennyweight (24 grs.)   - = 1·55517 „

APOTHECARIES' WEIGHT :
  1 Ounce (8 drms.)   -   - = 31·10350 Grams.

METRIC TO IMPERIAL.

LINEAR MEASURE :
  1 Metre (m.) = 39·37079 in. = $\begin{cases} 3\cdot28089917 \text{ Feet.} \\ 1\cdot09363306 \text{ Yards.} \end{cases}$

SQUARE MEASURE :
  1 Sq. Metre or Centiare (100 sq. decim.) - - $\left.\right\} = \begin{cases} 10\cdot76430 \text{ Sq. Feet.} \\ 1\cdot19603 \text{ Sq. Yard.} \end{cases}$
  Are (100 sq. m.)   -   - = 119·60333 Sq. Yards.

CUBIC MEASURE :
  1 Cubic Metre or Stere $\left.\right\}$ (1,000 c.d.) - = $\begin{cases} 35\cdot31658074 \text{ Cubic Feet.} \\ 1\cdot30802151 \text{ Cubic Yard.} \end{cases}$

MEASURE OF CAPACITY :
  1 Litre or 1,000 Cubic Centimetres or 1 Cub. Decimetre $\left.\right\}$ = 1·76077 Pints.

AVOIRDUPOIS WEIGHT :

1 Gram	-	-	- =	15·43235 Grains.
1 Kilogram (1,000 grams)		- =	$\left\{\begin{array}{l} 2\cdot20462125 \text{ Lbs. or} \\ 15432\cdot84874 \text{ Grains.} \end{array}\right.$	

TROY :

1 Gram	-	-	- =	$\left\{\begin{array}{l} 0\cdot03215073 \text{ Oz. Troy.} \\ 0\cdot64301 \text{ Pennyweight.} \\ 15\cdot43235 \text{ Grains.} \end{array}\right.$

With reference to the **above trade** equivalents it may be stated that the equivalent of the metre, 39·37079 inches, is based **on** a comparison made at Paris in 1818 by M. Arago, the Astronomer, and Capt. H. Kater, F.R.S. ("Phil. Trans. Roy. Soc." London, 1818), of two platinum metres for the Royal Society. These metres were made in consequence of a resolution of the House of Commons of 15th March 1816, moved by Mr. Davies Gilbert, **M.P.**, and one on 14th May 1817, by which the **Standard** Yard (Bird's, 1760), in the custody of the **Clerk of the** House, was required to be delivered to the Royal **Society,** so that it might be **compared "with the ten-millionth part of** the quadrant of **the meridian." Capt. Kater,** however, did not base his equivalent **on Bird's yard,** but on Sir G. Shuckburgh's subdivided standard scale of 1795, which was deposited with the Royal Society.

One of **the** earliest official attempts to define the English equivalents of Imperial weights and measures was that made at Paris by an officer of the Army **of** Occupation for the convenience of the individuals composing it, which calculations were published at Paris on **1st** January **1817, having been** printed at the headquarters of the army.

A direct comparison **of the** Imperial Yard with the new international standard metre has now been made under the directions of the Board of Trade and the International

Committee of Weights and Measures (see Board's Annual Reports, 1890 and 1895, and "Détermination du Rapport du Yard au Metre," par M. le Dr. J. René Benoit, Director du Bureau International des Poids et Mesures. Paris, 1896), with the following result :—

At 16°·667 Centigrade the **Imperial Yard** is equal to 0·9143992 metre, the temperature 16°·667 being taken as equivalent to 62° Fahrenheit; or, at 16°·667 C. (62° F.) the metre is equal to 39·370113 inches.

The above equivalent, 39·370113 inches, appears to confirm the deductions arrived at by Dr. Benoit in 1892 as to the value of the metre, viz., 39·36994 inches as deduced from the length of Bessel's Toise ("Etudes sur la Toise de Bessel et la Toise du Perou"), a value which was practically indentical with General Walker's comparisons of a Standard Yard of 1798 (Proceedings of the Royal Society, London, 1890). Dr. Gould in his recent annual address before the American Metrological Society has given 39·37000 inches, a mean value which was adopted by the United States Congress in 1866, when the permissive introduction of the metric system into that country was legalised. Mr. O. H. Tittman, in charge of the Weights and Measures Office at Washington, has found that a mean value, 39·37008 inches, agrees with values previously found within any uncertainty which the use of different thermometer scales may effect (United States Coast and Geodetic Survey Report, 1892). Professor W. A. Rogers has also found by experiment that 39·370155 inches represent the true relation between the lengths of the yard and the metre (Colby University, United States, March 29th, 1893).

Also the legal equivalent of the pound 453·59265 grams, was based on a comparison made at Paris in 1844 by Professor W. H. Miller, F.R.S., between a platinum kilogram which he had carefully adjusted, and the old kilogramme-des-archives. In 1883, however, a direct comparison was authorised to be made between the present Imperial avoirdupois pound and the new international " kilogramme prototype " (Board of Trade Report, 1884 : " Travaux et Mémoires Comité," Tome IV., 1885). As the result of the comparisons made in 1883 it appears that the pound is equal to 453·5924277 grams weighed in vacuô at 0° C. Based on the above determinations we have the following equivalents: the weight of water contained in the litre at 4° C. being compared with the weight contained in the gallon at 16°·667 C.

IMPERIAL TO METRIC (1896).

New scientific equivalents.

LINEAR MEASURE :

1 Inch	-	=	0·02539998	Metre =	25·399978 Millimetres.
1 Foot (12 in.)	-	=	0·3047997	,,	
1 YARD (3 ft.)	-	=	0·9143992	,,	
1 Pole (5½ yd.)	-	=	5·0291956	,,	
1 Chain (22 yd. or 100 links)	-	=	20·116782	,,	
1 Furlong (220 yd.)	=		201·1678	,,	
1 Mile (1760 yd.)		=	1609·34259	,, =	1·6093426 Kilometres.

SQUARE MEASURE :

1 Square Inch	-	=	0·0006451589	Sq. Metre =	6·451589 Sq. Centm.
1 Sq. Ft. (144 sq. in.)	=		0·0929029	,,	
1 Sq. YARD (9 sq. ft.)	=		0·836126	,,	
1 Sq. Perch (30¼ sq. yd.)	-	=	25·29281	,,	
1 Rood (40 perches)	=		1011·7123	,,	
1 ACRE (4840 sq. yd.)	=		404·6849	,,	= 0·40468 Hectare.
1 Sq. Mile (640 acres)	=		2589983·6	,,	= 258·99836 Hectares.

CUBIC MEASURE :

1 Cub. Inch	-	-	= 0·00001638702	Cub. Metre =	16·38702 Cub. Centms.
1 Cub. Foot (1728 cub. in.)	-		= 0·02831677	,,	
1 Cub. YARD (27 cub. ft.)	-		= 0·76455285	,,	

APOTHECARIES' MEASURE :

1 Gallon (8 pints or 160 fluid ounces) - - - =	4·54596 Litres.
1 Fluid Ounce (8 drachms) - =	2·84123 Centilitres.
1 Fluid Drachm (60 minims) - =	3·55153 Millilitres.

NOTE.—The Apothecaries' gallon is of the same capacity as the Imperial gallon.

MEASURE OF CAPACITY:

1 Gill - - - - =	1·42061 Decilitres.
1 Pint (4 gills) - - =	0·56825 Litre.
1 Quart (2 pints) - - =	1·13649 Litres.
1 GALLON (4 quarts) - - =	4·5459631 ,,
1 Peck (2 galls.) - - =	9·09193 ,,
1 Bushel (8 galls.) - - =	3·63677 Decalitres.
1 Quarter (8 bushels) - - =	2·90942 Hectolitres.

AVOIRDUPOIS WEIGHT :

1 Grain - - - =	64·79891824 Milligrams.
1 Dram - - - - =	1·77185 Grams.
1 Ounce (16 dr.) - - =	28·34953 ,,
1 POUND (16 oz, or 7,000 grains) =	0·45359243 Kilogram.
1 Stone (14 lb.) - - =	6·35029 ,,
1 Quarter (28 lb.) - - =	12·70059 ,,
1 Hundredweight (112 lb.) - =	{ 50·80235 Kilograms, or 0·50802 Quintal.
1 Ton (20 cwt.) - - =	1·01604704 Millier or Tonne.

TROY WEIGHT :

1 Troy OUNCE (480 grains) - =	31·10348 Grams.

NOTE.—The Troy grain is of the same weight as the Avoirdupois grain.

APOTHECARIES' WEIGHT :

1 Ounce (8 drachms) - - =	31·10348 Grams.
1 Drachm (3 scruples) - - =	3·88794 ,,
1 Scruple (20 grains) - - =	1·29598 ,,

METRIC TO IMPERIAL (1896).

LINEAR MEASURE:

1 micron (μ) - - =	0·001 mm.
1 millimetre (mm.) - =	0·001 m.
1 centimetre (cm.) - =	0·01 m.
1 decimetre (dm.) =	0·1 m.

1 **Metre (m.)** - - - $= \begin{cases} 39\cdot370113 \text{ Inches.} \\ 3\cdot2808428 \text{ Feet.} \\ 1\cdot09361426 \text{ Yard.} \end{cases}$

1 Decametre **(dam.)** - = 10 m.
1 Hectometre **(hm.)** - = 100 m.
1 Kilometre (km.) - = 1,000 m.
1 Myriametre - - = 10,000 m. - = 6·21371 Miles.

SQUARE MEASURE:
1 Sq. Centimetre (cm².) = 0·0001 m².
1 Sq. Decimetre (dm².) = 0·01 m².

1 Sq. **Metre (m².)** - - - $= \begin{cases} 1550\cdot006 \text{ Sq. Inches.} \\ 10\cdot763929 \text{ Sq. Feet.} \\ 1\cdot1959921 \text{ Sq. Yard.} \end{cases}$

1 Sq. Decametre (dam².) = 100 m².
1 Sq. Hectometre **(hm²)** or 100 Ares . } = 10,000 m² = 2·471058 Acres.

CUBIC MEASURE:
1 Cub. Centimetre (cm³.) = 0·000001 m³.
1 Cub. Decimetre (dm³.) = 0·001 m³.

1 **Cub. Metre (m³,)** - - - $= \begin{cases} 61023\cdot90 \text{ Cub. Inches.} \\ 35\cdot3147589 \text{ Cub. Feet.} \\ 1\cdot3079540 \text{ Cub. } \textbf{Yard.} \end{cases}$

MEASURE OF CAPACITY:
1 Microlitre - - - = 0·001 Millilitre.

1 Millilitre (ml.) ($\frac{1}{1000}$ litre) - = 0·00704 Gill.
1 Centilitre ($\frac{1}{100}$ „) - = 0·07039 „
1 **Decilitre** ($\frac{1}{10}$ „) - = 0·17598 Pint.
1 **Litre** - - = 1·75980 Pints.
1 Decalitre (10 litres) - - = 2·19975 Gallons.
1 Hectolitre (100 „) - - = 2·74969 Bushels.
1 Kilolitre (1,000 „) - - = 3·43712 Quarters.

WEIGHT: AVOIRDUPOIS.

1 Milligram (mgr.) - - = 0·01543 Grain.]
1 Centigram ($\frac{1}{100}$ gram.) - = 0·15432 „
1 Decigram ($\frac{1}{10}$ „) - = 1·54324 Grains.
1 **Gram** - - = 15·43236 „
1 Decagram (10 grams.) - = 5·64383 Drams.
1 Hectogram (100 „) - = 3·52740 Ounces.
1 **Kilogram (1,000 „)** - $= \begin{cases} 2\cdot20462234 \text{ Lb. or} \\ 15432\cdot35639 \text{ Grains.} \end{cases}$
1 Myriagram (10 kilog.) - = 22·04622 Lb.
1 Quintal (100 „) - = 1·96841 Cwt.
1 Millier or Tonne (1,000 kilog.) = 0·98420640 Ton.

TROY.

$$1 \text{ GRAM} = \begin{cases} 0\cdot03215074 \text{ Oz. Troy.} \\ 15\cdot43236 \text{ Grains.} \end{cases}$$

APOTHECARIES.

$$1 \text{ GRAM} = \begin{cases} 0\cdot25721 \text{ Drachm.} \\ 0\cdot77162 \text{ Scruple.} \\ 15\cdot43236 \text{ Grains.} \end{cases}$$

For many of the common purposes of retail trade it would appear to be possible to use abbreviated equivalents, such as follows :—

Abbreviated Metric Equivalents and Denominations.

ABBREVIATED EQUIVALENTS.

1 millimetre	$= \frac{4}{100}$ of an inch.
1 metre	$=$ 3 feet $3\frac{1}{2}$ inches.
1 kilometre	$= \frac{6}{10}$ of a mile.
1 square metre	$= 1\frac{2}{10}$ square yard.
1 cubic metre	$= 1\frac{3}{10}$ cubic yard.
1 litre	$= 1\frac{3}{4}$ pint.
1 gram	$= 15\frac{4}{10}$ grains.
1 kilogram	$= 2\frac{1}{5}$ lb. avoir. or $2\cdot2$ lb.
1 tonne or millier	$=$ ton.

1 inch	$= 25\frac{4}{10}$ millimetre.
1 foot	$= \frac{3}{10}$ of a metre.
1 yard	$= \frac{9}{10}$ of a metre.
1 mile	$= 1\frac{6}{10}$ kilometre.
1 square yard	$= \frac{8}{10}$ square metre.
1 acre	$= \frac{4}{10}$ of the hectare.
1 cubic yard	$= \frac{3}{4}$ of a cubic metre.
1 gallon	$= 4\frac{1}{2}$ litres.
1 quart	$= 1\frac{1}{10}$ litre.
1 pint	$=$ nearly $\frac{1}{2}$ litre.
1 pound	$= 0\cdot45$ kilo or nearly $\frac{1}{2}$ kilo.
1 cwt.	$= \frac{1}{2}$ quintal.
1 ton	$=$ 1 tonne.

The following are the recognised international symbols of the legal denominations **of metric** weights **and** measures :—

Mesures de longueur.	Symbole.	Mesures de surface.	Symbole.	Mesures de volume.	Symbole.
Kilomètre - -	km	Kilomètre carré	km²	Mètre cube -	m³
Mètre - -	m	Hectare - -	ha	Stère - -	s
Décimètre - -	dm	Are - -	a	Décimètre cube	dm³
Centimètre -	cm	Mètre carré -	m²	Centimètre cube	cm³
Millimètre - -	mm	Décimètre carré	dm²	Millimètre cube	mm³
Micron - -	μ	Centimètre carré	cm²	- - -	—
= 0·001 mm -		Millimètre carré	mm²	- - -	—

(*Continued.*)

Measures de capacité.	Symbole.	Poids.	Symbole.
Hectolitre -	hl	Tonne - -	t
Décalitre - -	dal	Quintal métrique	q
Litre -	l	Kilogramme -	kg
Décilitre -	dl	Gramme -	g
Centilitre -	cl	Décigramme -	dg
Millilitre - -	ml	Centigramme -	cg
= 0·001 l.		Milligramme -	mg
		Microgramme -	y
		= 0·001 mg.	

For instance, 100m means 100 metres ; or 100m² = 100 square metres ; or 1000g = 1000 grams, &c.

The Metric system had its origin in 1790, when an attempt was made by the French National Assembly (Decree, 8th May 1790) to make the metric system compulsory in France, and to establish a decimal system based on the metre, or the one ten-millionth part of the quadrant of a meridian (*see* " Base du Système Métrique Décimal ou mesure de l'arc du méridien entre Dunkerque et Barcelone " par MM. Méchain et Delambre, Paris. 1810.); the gram

Original metric standards.

weight then being declared to be equal to the weight of
distilled water contained in a cubic centimetre (or a vessel
whose linear dimensions were equal to one-hundredth part
of the length of the metre). The original standards were
made in 1795, 18 Germinal, An. III. (7th April, 1795),
under the direction of the National Institute of Science,
and in 1799, 19 Frimaire, An. VIII. (9th December, 1799),
were deposited at the Palais des Archives at Paris. In the
years 1801 and 1812 (Decrees, 3rd November 1800 and
12th February 1812), however, the French Government
found it necessary to again permit the use in trade of the
old French weights and measures (système-usuel) and it
was not until 1840 (Law, 4th July 1837) that their use
was finally forbidden. In other countries, however, the
metric system has been made legally compulsory in less
time—from two to ten years. Some of the old French
denominations were similar to ours, as pinte or pint,
tonneau or ton, boisseau or bushel, once or ounce, grain,
mille, perch, &c.

If the Standard Metre was lost or injured it would not
now be restored by reference to the meridian, but by
reference to existing material copies of the present metre
prototype deposited with the various high contracting
States who have joined the Metric Convention of 1871.
The ideal metre (Clarke's "Geodesy," 1880), as determined
by more recent geodetic measurements, is not equivalent to
39·37079 inches, but to 39·37779, as compared with Clarke's
value of the metre, viz., 39·37042 inches, not a difference
(0·007 inch) of much trade importance, perhaps, but one
of importance for scientific and manufacturing purposes.

The gram also would be restored by reference to the
"kilogram-prototype" at Paris (1,000 grams), and not
with reference to the weight of water. The capacity of the
litre measure, similarly, would not depend on the weight
of a cubic decimetre of water, but would be taken as equal

to a kilogram weight of water. In hydrostatic weighings densities and volumes should be expressed therefore in litres or parts of a litre, and not in cubic centimetres; for instance, in a millilitre (ml.) or a gram weight of water at 4° C., and not in cubic centimetres.

The actual weights and measures used in workshops and factories are represented by the following series :—

Denomination of metric weights.

Kilograms.	Grams.	Decigrams.
500	500	5
200	200	2
200	200	2
100	100	1
50	50	Centigrams.
20	20	5
20	20	2
10	10	2
5	5	1
2	2	Milligrams.
2	2	5
1	1	2
		2
		1

Iron and brass weights.

Double decalitre, or	-	-	20	litres.
Decalitre	-	-	10	„
Double litre	-	-	2	„
Litre	-	-	1	„
Demi-litre	-	-	0·5	„
Double decilitre	-	-	0·2	„
Decilitre	-	-	0·1	„
Demi-decilitre	-	-	0·05	„
Double centilitre	-	-	0·02	„
Centilitre	-	-	0·01	„

Measures of capacity made of brass, iron, wood, &c.

A glass litre measure, sub-divided into 1,000, 500, 200, 100, 50, 20, 10, 5, 2, 1 millilitres (ml.).

Metre, sub-divided into 10 decimetres, the decimetre being divided into 10 centimetres and the centimetre into 10 millimetres. Double metre, 20-metre, and 40-metre chains.

Measures of length made of brass, steel, wood, &c.

It will be seen that in the above convenient series 5, 2, 2, 1, there is a duplicate weight of the 2, so as to obtain the multiple 10. A metric series is also in use abroad, based on weights of the proportions 5, 2, 1, 1 ; our decimal troy ounce series is 5, 4, 3, 2, 1 ; thus in the former series a third weight, 1, is needed to make up 10, making altogether five weights in place of the four used in the series 5, 2, 2, 1. In some grain weight series a convenient duodecimal series is also followed, as 12, 6, 3, 2, and 1.

The facing diagrams (Figures 22, 23, and 24) show the shapes of the ordinary metric weights and measures.

Teaching of metric weights and measures in schools. The proper way in which weights and measures may be taught in schools is shown in the Regulations of the Committee of the Privy Council on Education as hereafter referred to; but arithmetic books, unfortunately, often contain numerous local and customary weights and measures and obsolete terms of no practical use to a student. For instance, in a well known modern school book we are informed that cloth is measured by the "ell," a measure which legally and locally expired in this country half a century since, and a distinction is even drawn between the English and the obsolete Flemish ell. The student is further told that the hide is a measure of land although the hide was obsolete in 1701, and so on.

Probably the only weights and measures that need be taught, are as follows :—

IMPERIAL SYSTEM.   METRIC SYSTEM.

Weights.

16 drams = 1 ounce (oz.)	1 kilo-gram = 1,000 grams.
16 ounces = 1 pound (lb.)	1 hecto-gram = 100 ,,
14 pounds = 1 stone.	1 deca-gram = 10 ,,
8 stones = 1 hundred-	1 GRAM.
weight (cwt.)	1 deci-gram = 0·1 gram.
20 cwt. = 1 ton.	1 centi-gram = 0·01 ,,
1 *lb. avoirdupois also con-*	1 milli-gram = 0·001 ,,
tains 7,000 grains.	1 *kilo-gram equals* 2²⁄₁₀, *or*
1 *ounce troy* = 480 *grains*	2·2 *avoirdupois pounds*
avoirdupois.	*nearly.*

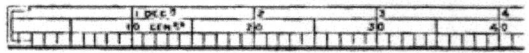

DECIMETRE

DRAWN ⅓ SCALE

BRAS

DRAWN ⅓ SCALE

Fig. 22.—IRON WEIGHTS, ENGLISH PATT

DRAWN $\frac{5}{6}$ SCALE

DOUBLE DECIMETRE

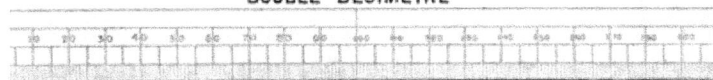

DRAWN $\frac{3}{5}$ SCALE

AND WOODEN MEASURES.

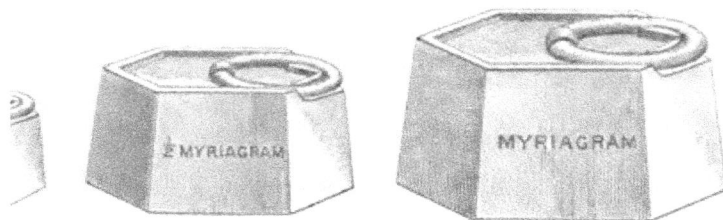

2 MYRIAGRAM

MYRIAGRAM

ERN, BUT FRENCH SERIES.

DRAWN TO ¼ SCALE

DECIGRAMS

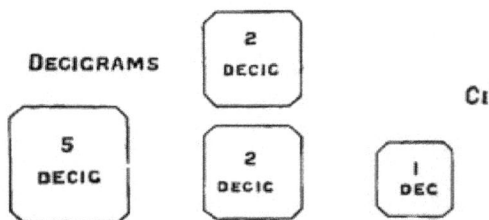

Fig. 23.—BRASS WEIGHTS

2
LOG^{MS}

2
KILOG^{MS}

5
KILOGRAMS

10
KILOGRAMS

2
C.G

NTIGRAMS

MILLIGRAMS

5
C.G

2
C.C

1
C.G

5
M

2
M

2
M

1
M

DRAWN FULL SIZE

ENGLISH PATTERN AND SERIES.

(BRASS OR PEWTER

DRAWN

Fig. 24.—METRIC MEASURES OF CAPACITY

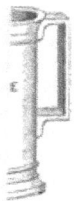   

MEASURES).

SCALE

(WOODEN MEASURES).

Linear Measure.

12 inches	= 1 foot.	1 kilo-metre = 1,000 metres.	
3 feet	= 1 yard.	**1 Metre.**	
5½ yards	= 1 pole.	1 deci-metre = 0·1 metre.	
40 poles	= 1 furlong.	1 centi metre = 0·01 „	
8 furlongs	= 1 mile.	1 milli-metre = 0·001 „	

1 *metre* **equals** 3 *feet* 3½ *inches.*

Measures of Surface.

144 square inches = 1 square foot.	1 **are**	= 100 square metres.
9 „ feet = 1 square yard.	1 **centi-are**	= 1 square metre.
4,840 „ **yards** = 1 acre.		

Measures of Volume.

1,728 cubic inches = 1 cubic foot.	1 **stere** = 1 cubic **metre.**
27 cubic feet = 1 cubic yard.	

Measures of Capacity.

4 gills	= 1 pint.	1 hecto-litre = 100-litres.
2 pints	= 1 quart.	1 deca-litre = 10 „
4 quarts	= 1 gallon.	**1 Litre.**
2 gallons	= 1 peck.	1 deci-litre = 0·1 litre.
4 pecks	= 1 bushel.	1 centi-litre = 0·01 „

1 *gallon* **of** *pure* **water** *weighs* **10 lbs.** *avoirdupois.*

1 *litre of pure water weighs* 1 *kilogram.*

The Committee of the Privy Council on Education in a Memorandum (Mem. 5) issued **in August 1893,** to their Inspectors on Examination in the metric system, **remark** that the education code provides that scholars in the fifth and higher standards shall be examined **in** the principles of the metric system, *i.e.,* **on the** convenience of adopting systems **of** coinage **and of** weights and measures in which the increase of values or of quantities proceeds by multiples of **ten, and** their diminution by tenths. So long as these

principles are grasped, it is immaterial whether instruction
in the system is illustrated by the metre, with its sub-
divisions and multiples, or by some other unit; but it
will probably be found most convenient to illustrate them
by reference to the metre and the franc. The metre can
be readily compared with the yard, and its approximate
length can be readily remembered, if it is taught in the
form of 3 feet 3½ inches, in which the number three alone
is employed. No other calculation should be required
than small addition sums, showing that every complete
ten of a lower denomination forms a complete unit of
the next higher denomination; e.g.—

$$fr. \qquad fr. \qquad fr. \qquad fr.$$
$$3\cdot75 \; + \; 5\cdot25 \; + \; 4\cdot5 \; = \; 13\cdot50.$$

One school inspector in his report suggested that
" School managers would do well to adopt an old sugges-
" tion, and provide their schools with such of the standard
" weights and measures as they conveniently can; a few
" hours weighing and measuring would make the children
" more at home with them than many hours of bare
" learning the tables", and some School Boards now
provide educational models of weights and measures; such
as those designed by Mr. F. W. Levander, F R.A.S., and
Mr. G. Evans.

PART VI.

15.—WEIGHTS AND MEASURES USED FOR SPECIAL PURPOSES.

The weights and measures used by pharmacists, or chemists and druggists, are regulated for the purposes of trade by the Weights and Measures Act, 1878, s. 20, which provides that "Drugs, when sold retail, may be sold by apothecaries' weight." For dispensing purposes, the Medical Act, 1858, section **54**, provides that the General Council of Medical Education and Registration of the United Kingdom "shall cause to be published, under their " direction, a book containing a list of medicines and " compounds, together with the true weights and measures " of which they are to be mixed." This published book is known as the "British Pharmacopœia," which by the Act 25 and 26 Vict. c. 91. (1862) superseded the Pharmacopœias hitherto in use in England, Scotland, and Ireland. In the British Pharmacopœia of 1885, the edition still in use, both Imperial and Metric weights and measures are used, Imperial measures only in the body of the book, and the alternative use of metric weights and measures is extended to analysis, whether gravimetric or volumetric. In the new forthcoming edition of the Pharmacopœia it is understood that this arrangement will be modified by a more extensive application of the metric system, The avoirdupois ounce

Pharmaceutical or apothecaries' weights and measures.

and pound are the weights practically used in the sale of medicines :

Weights.

1 grain (*gr.*)
1 ounce (*oz.*)   -   -   = 437·5 grains.
1 pound (*lb.*)   -   -   = 16 ozs. = 7,000 grains.

Measures.

1 minim (min.)
1 fluid drachm (fl. dr.)   -   = 60 minims.
1 fluid ounce (fl. oz.)   -   = 8 fluid drachms.
1 pint (O.)   -   -   -   = 20 fluid ounces.
1 gallon (C.)   -   -   -   = 8 pints.

Relation of Measures to Weights.

1 minim (m.) is the measure of 0·91 grain weight of water.
1 fluid drachm (*f.* ʒ)   -   54·68   ,,
1 fluid ounce (*f.* ʒ)   -   437·5   ,,
1 pint (O.)   -   ·   8,750·0   ,,
1 gallon   -   -   70,000·0   ,,

Apothecaries' (Old) Weight.

Mark.
20 grains (*gr.*)   -   -   = 1 scruple,   - Ɔ
3 scruples, *or* 60 grains   = 1 drachm,   - ʒ
8 drachms, *or* 480 grains   = 1 ounce,   - ʒ
12 ounces, *or* 5,760 grains   = 1 pound,   - *lb.*

The scruple and drachm, although not introduced into the Pharmacopœia of 1885, are still used in prescriptions.

The Fahrenheit scale of temperature is followed and specific gravities are referred to a standard temperature of 60° F. The Pharmacopœia recognises that graduated glass measures of Imperial denominations should be graduated at the legal temperature of 62°, the common practice with chemists, but, with regard to volumetric vessels and specific gravity bottles, these have to be marked at 60° F., a temperature also commonly recognised by chemists for volumetric purposes. If, therefore, volumetric measures are used graduated at 60° in conjunction with Imperial measures, the determination would require a slight correction if exact results are to be obtained.

The several denominations of apothecaries' weights and measures legal for use in trade are given in Appendix No. 1 ; such denominations were legalised by H.M.'s Order in Council, dated 14 August 1879. The Annual Calendar of the Pharmaceutical Society of London gives also a list of all the Imperial and Metric weights and measures used in pharmacy.

According to Hoblyn ("Dictionary of Medical Terms") the Edinburgh and Dublin Colleges formerly retained the old signs for the capacity measures—for the gallon *cong.* (Roman Congius); for the pint *lb.*; the ounce ℥; the drachm ℨ; scruple ℈, and the drop *gt.* (French goutte). Dr. Ch. R. C. Tichborne acquaints the writer that these old signs also are still in use in prescription writing, particularly the last four.

The following weights and liquid measures were set forth in the Pharmacopœia Collegii Regalis Medicorum, Londinensis, 1851, as those then authorised to be used by apothecaries :—

PONDERA :

Libra	- ℔			
Uncia -	- ℥		Uncias duodecim	- ℥ xij
Drachma	- ℨ	habet	Drachmas octo	- ℨ viij
Scrupulus	- ℈		Scrupulus tres	- ℈ iij
Granum	- gr :		Grana viginti	- gr : xx

MENSURÆ :

Conigus	- C			
Octarius	- O		Octarios Octo	- O viij
Fluid uncia -	f℥	habet	Fluid uncias viginti	f xx
Fluid drachma	fℨ		Fluid drachmas octo	fℨ viij
Minimum	- ♏		Minima sexaginta	- ♏ ix

I

Coin
weights.

In pursuance of the Coinage Act, 1870, standard weights are kept of the current coins of the realm of the following denominations. These standards are made of gold, palladium, silver, and bronze :—

Five sovereigns, two sovereigns, sovereign, half-sovereign, crown, half-crown, double-florin, florin, shilling, sixpence, groat or fourpence, three-pence, twopence, silver penny ; also bronze penny, half-penny, and farthing.

The Act specifies the weights of the coins, and provides that in making the coins a "remedy," or variation from true standard weight may be permitted. The standard weight of the sovereign and half-sovereign, for instance, is 123·27447 and 61·63723 grains respectively; and the least current weight is 122·5 and 61·125 grains respectively. The remedy on the sovereign and half-sovereign coins has, under the Coinage Act, 1891, been fixed at 0·2 and 0·15 grain, and the Act provides that a loss exceeding 3 grains should be primâ facie evidence that the coin has been impaired, diminished, or lightened otherwise than by fair wear and tear.

Coin weights, for bankers use are verified and stamped, if not less in weight than the weight of the lightest coin for the time being legally current, and small brass coin weights or brass discs representing the weight of the sovereign and half-sovereign so verified are just weights for determining the weight of gold and silver coin.

The following are the legal weights of the coins of the realm :—

Denomination of Coin.	True Standard Weight.	Least Current Weight.
	Grains.	Grains.
Gold :		
Five pound - - - -	616·37239	612·500
Two pound - - -	246·54895	245·000
Sovereign - - -	123·27447	122·500
Half-sovereign - - - -	61·63723	61·125

Denomination of Coin.	True Standard Weight.	Least Current Weight.
	Grains.	
Silver :		
Crown - - -	436·36363	
Half-crown - - -	218·18181	
Florin - - -	174·54545	
Shilling - - -	87·27272	
Sixpence - - -	43·63636	
Groat or fourpence - -	29·09090	No least current
Threepence - - -	21·81818	weights for
Twopence - - -	14·54545	silver or
Penny - - -	7·27272	bronze coins.
Bronze		
Penny - - -	145·83333	
Half-penny - - -	87·50000	
Farthing - - -	43·75000	

Coin weights are of ancient origin, as is shown by a catalogue of Arabic glass weights in the British Museum prepared by Mr S. Lane-Poole. Weights in the form of small glass discs were used by the Arabs from the year A.D. 717 in testing the weights of coins; and such discs had inscribed on them their denominations as:—

" Weight of deenar " (65 grains.)
" Weight of half-deenar."
" Full weight."
" Weight of a fels of twenty keerats " (56 grains.)

Benedictory phrases are sometimes to be found on such weights, as :—

" In God's name."
" God's measure."
" God ordains honesty."

Closely allied to stamped glass weights are the curious glass verification stamps, glass discs, or " bottle-stamps " which were placed (A.D. 65 to 132) on Egyptian glass measures of capacity of the form of a bottle.

There was formerly at the English Royal Mint an officer called a " Weigher or Stamper of Money Weights,"

i 2

whose duty it was to test all coin weights, which duty has
since, by the Acts 33 and 34 Vict. c. 10, and 41 & 42 Vict.
c. 49. s. 39, passed to the Standards Department.

In connection with the standard coin weights, reference
may here be made to the Pyx Chapel or Chamber, at
Westminster Abbey, a depository for standards since the
Norman period.

In ancient times, and probably since the Norman period,
the Chamberlains of the Exchequer had the custody of the
" King's Standards," which at one time were kept in the
Pyx Chapel at Westminster Abbey.

Of the Pyx Chapel, Dean Stanley has given an
interesting description in his " Memorials of Westminster
Abbey." In the Eastern Cloister is an ancient double door
made of oak, each door having three locks, which admits
to the Pyx Chapel or Chamber, a building belonging to
the Norman substructure underneath the dormitory, and
which was " no less than the Treasury of England, a
" grand word, which, whilst it conveys us back to the
" most primitive times, is yet big with the destinies of the
" present and the future of that sacred building, in which
" were hoarded the treasures of the nation."   *   *   *
" It was probably almost immediately after the Conquest
" that the Kings determined to lodge their treasure 'under
" the guardianship of the 'inviolable sanctuary' which
" 'St. Peter had consecrated, and the bones of the
" 'Confessor had sanctified.' "

Two views of the interior of the Pyx Chamber are given
in Figures 25 and 26. The altar shown therein is of later
date than the chapel, and it appears that at one time it
may have been used by the King's moneyers for Coinage
purposes. On the right and left of the pillar shown in
Figure 26, which stands in the middle of the chamber

Fɪɢ. 25.

Fɪɢ. 26.

INTERIOR OF THE PYX CHAPEL AT WESTMINSTER.

there are a number of old oak presses (temp. 1610) in which State Records were formerly kept. Behind the pillar or on the east side of the chamber there are also two large " treaty " chests ; and, besides these, there are two smaller chests in which standards and " assays of gold and silver " were kept, of which ancient chests two illustrations are given in Figure 27.

A view of the entrance to the Chapel from the eastern cloisters of Westminster Abbey (near the Chapter House) is given in the frontispiece. The massive double doors therein shown are of later date than the chapel and were formerly secured by six locks, three on each door, each set of keys being in the custody of a separate officer ; a sketch of the old keys, which are still in use, is given in Figure 28.

The " Trial of the Pyx," is now held annually at the Goldsmiths' Hall, when an examination takes place of the justness of the gold and silver coins of the realm as issued for the past year ; it probably took its name from the Pyx Chapel, and is an ancient ceremony practised in the country probably so early as in Saxon times.

By the Coinage Act, 1870, it is provided that the standard weights for testing the coin of the realm shall be placed in the custody of the Board of Trade. It is also provided by the Act that the Master of the Mint shall, from time to time, cause copies to be made of such standard weights, and that once at least in every year the Board of Trade and the Master of the Mint shall cause such copies to be compared and duly verified with the standard weights in the custody of the Board of Trade.

By the same Act it is also provided that the standard trial plates of gold and silver used for determining the justness of the gold and silver coins of the realm shall be in the custody of the Board of Trade, and that the performance of all duties in relation to such trial plates shall be part of the business of the Standards Department. In 1873 new trial plates of pure gold and pure silver (millesimal fineness equals 1,000) and of standard gold and standard silver (916·66 and 925) were made and copies of such plates were issued to the Royal Mint, and also to the Indian and Australian Mints, as well as to the Assay Offices at Birmingham, Sheffield, Edinburgh, Dublin, Exeter, Newcastle, Chester, and Glasgow.

Regulations respecting the trial of the pyx, and the production before a jury of the Goldsmiths' Company of London of the standard trial plates and standard weights were made by an Order in Council dated 29th June 1871.

The Jewel Tower at Westminster.

Another ancient depository for standards and coin weights is the Jewel Tower or Jewel House on the north-east corner of the Dean's Garden.

The Jewel Tower is stated to have been built in 1350 (temp. Richard II.), and to have been designed by Abbot Litlington. It appears to have been sold by the Abbey to the Crown in the last year of Edward III., and was at one time used as a temporary treasury for the King's jewels, and subsequently became a depository of the Acts of Parliament and of the Peers' records. For a little while it formed part of the residence of the Speaker of the House of Commons, during the rebuilding of the Houses of Parliament after the fire of 1834, and in 1866 came into the custody of the Board of Trade. In this tower Dean Stanley states that the long lost Prayer Book of 1662 was found, which had been detached from the Act of Uniformity, and had lain hid in the Parliament Office at the Jewel Tower.

This tower has three floors, on each of which is a large chamber and a small ante-room the upper floor being reached by a winding flight of stone steps, and the only

FIG. 27.—TWO ANCIENT CHESTS IN WHICH STANDARDS, &c.,
WERE FORMERLY KEPT IN THE PYX CHAPEL.

(See page 121.)

approach to the Tower is through Old Palace Yard. A view of the basement chamber of the Tower is given (by permission of Messrs. Cassell) in Fig. 29 (see also Fig. 18).

Fig. 29.— BASEMENT OF THE JEWEL TOWER NOW USED AS A BALANCE ROOM.

Although multiple weights of the sovereign, as 500, 200, 100, and 50 sovereigns are used by bankers, no standards of such weights have been found to be practically necessary, as they are not used for determining the actual weight of coin, but only as a ready means of ascertaining the tale, up to 100 or 200 sovereigns, without counting piece by piece. A multiple weight representing 100 sovereigns least current weight could not be used to prove that 100 sovereigns

Bankers' weights.

placed in the opposite pan of the balance were all and each of them legal current coins or the contrary. The average weight of gold coins, as adopted by London scale-makers for adjusting bankers' weights, was for some years taken as 12289 grains to the 100-sovereign weight ; but recently, since the withdrawal of the light gold coin, the weight taken has been 12300 grains for the 100-sovereign weight, or 25 ozs. : 12½ dwts. troy.

It is, of course, even more difficult to state the average current weight of silver coins, but some scale-makers take as bankers' weight for silver :—10l.=35 ozs. and 4 dwts. troy.

Besides " coin " weights, bankers and assayers use the following decimal series of troy weights introduced in 1853 (Act for regulating the Weights used in the Sale of Bullion), 500, 400, 300, 200, 100 ounces troy, proceeding by a similar decimal series down to 0·001 ounce. The old troy series, 12, 6, 3, 2, 1 ounce is no longer used at the Bank of England.

Weights used by silver-smiths and pawn-brokers.

The weights used by silversmiths and pawnbrokers are generally troy ounces and pennyweights (dwts.), or grain weights as follows : The Weights and Measures Act, 1878, s. 20, provides gold and silver, and articles made thereof, including gold and silver thread, lace or fringe, also platinum, diamonds, and other precious metals or stones, may be sold by the ounce, troy, or by any decimal parts of such ounce, and all contracts, bargains, sales, and dealings in relation thereto shall be deemed to be made and had by such weight, and when so made and had shall be valid :—

480 grains or 20 dwts. = 1 ounce troy.
240   ,,   = 10 pennyweights.
120   ,,   = 5   ,.
 72   ,,   = 3   ..
 48   ,,   = 2   ,,
 24   ,,   = 1   ,,

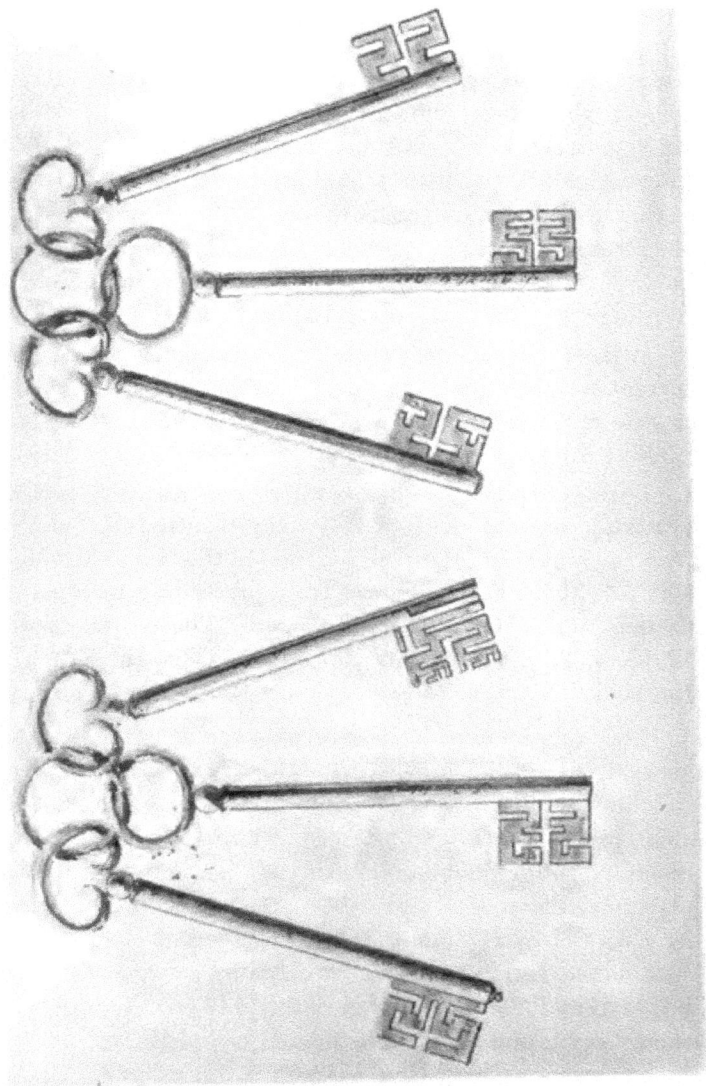

Fig. 28.—ANCIENT KEYS OF THE TWO DOORS OF THE PYX CHAPEL.

(See page 121.)

Besides the decimal bullion weights (500 to 0·001 ounce) the old binary series of troy weights is used by some silversmiths :—

> 10, 8, 6, 4, 2, 1 ounce, troy.
> 10, 5, 3, 2, 1 pennyweight.

The pound troy of 12 troy ounces (5,670 grains) is no longer permitted for use in trade.

In the assay of the precious metals there is still used the ancient "carat" weight. The *gold* carat is an imaginary weight; the assayer estimates that pure gold is divided into 24 parts, called carats, each carat being also subdivided into 24 parts, called carat-grains, and formerly each carat-grain again into eight parts. Thus standard gold, of which wedding-rings are made, should contain 22 carats of gold and two carats of alloy (silver or copper). At the present day the gold carat is mostly used by retail goldsmiths and jewellers, but all the principal assayers follow the Royal Mint and the Royal School of Mines, and adopt the millesimal system of gold assay, in which pure gold is divided into 1,000 parts; thus a wedding ring of 22 carats would be 916·66 fine, the standard fineness of the sovereign and half sovereign coins.

For assay purposes an old assay pound of 12 grains was formerly much used. The following is "Refiners' weight," a weight which appears to be identical with what is known as the trade system of reporting gold assays, and to which many refiners still work (Chaffers, "Hall Marks on Gold and Silver Plate," London, 1875, p. 47) :—

> 12 ounces make 24 carats.
> 4 grains „ 1 carat.
> 4 quarters „ 1 grain.
> 15 grains troy make ½ grain.

Until 1882 gold assays were reported on that system at the Royal Mint; 1 carat pound being taken as equivalent to 24 carats or 5,760 grains.

And for assay troy weight the following ancient series is sometimes recognised :—

    12 ounces make 1 lb. troy, or 5,760 grains.
    20 pennyweights make 1 ounce (480 grains.)
    24 grains   „   1 pennyweight.
    20 mites   „   1 grain.
    24 droits   „   1 mite.
    20 periots   „   1 droit.
    24 blanks   „   1 periot.

Carat and Assay weights. According to Dr. Ure ("Dictionary of Arts, 1843, Art. "Diamond," p. 391), the term "carat" was said to be the name of a red bean which grew in the district of Shangallas, in Africa, once a famous mart for gold dust; and became transported into the East, where it was long employed in weighing diamonds; Morin, however, derives the word from an Arabic word "kyrat," signifying weight, and thinks this may be again derived from a Greek word "keration," a small weight. Mr. E. W. Streeter ("Precious Stones and Gems," 1877) states that the diamond carat is not of the same weight in all countries. There is no standard of this carat, but in England it may generally be taken as equal to nearly 205·409 milligrams; the ounce troy of 480 grains being sometimes divided into 151½ diamond carats also.

Sale of bread by weight. The sale of bread by weight, or the assize of bread, has been regulated by laws from the most ancient period, and probably before the Assiza Panis et Cervisiæ of 51 Henry 3. c. 18 (1266).

In the Mosaic law even we find reference to the delivery of bread by weight, as Lev. xxvi. 26, "They shall deliver you your bread again by weight"; and Ezekiel iv., 16 "they shall eat bread by weight and with care."

All bread, except fancy bread, is required to be sold by weight; and the sale of bread by weight, and the weights and weighing instruments to be used in the sale thereof, are mainly regulated by the Act 6 and 7 Will. 4. c. 37. (1836) ss. 3 to 7.

In Ireland the sale is regulated by the Act 1 and 2 Vict. c. 28 (1837–8), which repealed the Acts then in force for preventing the Adulteration of Meal, Flour, and Bread. The sections 3 to 6 of the Irish Act are similar in effect to those of the English Act.

In Scotland the Burgh Police (Scotland) Act, 1892, contains provisions with reference to the sale of bread in burghs. One provision of this Act enacts that all bakers and dealers in bread shall on all bread made or exposed by them for sale (except fancy bread or rolls) impress thereon in large and distinct figures the imperial weight of such bread.

In London the local Act 3 Geo. 4. c. 106 (1822), regulates the sale of bread also within the City of London and the liberties thereof within the weekly bills of mortality and within ten miles of the Royal Exchange.

Part III of the Weights and Measures Act, 1889 (s. 32), also explains the law as to bakers (3 Geo. 4. c. 106., s. 9 ; and 6 and 7 Will. 4. c. 37. s. 7).

We have already shown (Part 1, page 24) the intimate Sale of connection which existed between the standards of measure corn. and weight and wheat and barley, a connection, however, that has long ceased to exist. It is probable, that the present local practice as to the mode of selling grain is not very different from what it was a century ago, excepting as to the denominations of the measures. In the United Kingdom grain is sold (1) by measure (as by the Imperial Bushel); (2) by weight (by so many pounds); or (3) by a weighed measure (as by so many pounds to the bushel). This latter may be described as a qualitative measurement, the two first being purely quantitative.

In Ireland, the practice is largely to sell by weight only ; owing perhaps to the effects of the legislation of 1705 (4 Anne), when in Ireland all meal and flour was required to be sold by weight only, which in 1733 (7 Geo. 2. c 15) was extended to all corn sold in Ireland.

The denominations of the measures and weights generally used are household words, the peck, bushel, quarter, pound, stone, and hundredweight, all of which are Imperial denominations, but local and customary measures still flourish on some soils in the United Kingdom. In Scotland we find old Scotch names still surviving—as boll (240 lbs.), forpit &c. ; in East Anglia too we find the coomb of four Imperial bushels in use for barley and oats ; and in Wales the " hobbett " of wheat (hobaid or hobed) of four bushels, and sometimes 168 lbs. In Ireland, as well, we find sales also by the quarter of 480 lbs, or barrels of 20 stones, in the sale of wheat, and a barrel of 16 stones in the sale of barley ; and for oats a barrel of 14 stones. In 1879 there was legalised by Order in Council a new denomination of weight called the " cental," or 100 lbs., which is also sometimes used, particularly in the American grain trade.

Heaped measure is no longer legal, as it was finally abolished by the Weights and Measures Acts, 1878, Section 17, and 1889, Section 5.

The practice as to the sale of fruit and vegetables is not so uniform as in the sale of grain, and varies with locality and custom, and sometimes we find the name of the vessel in which the fruit, &c. is packed becomes recognised as and confused with a measure, as barrel, basket, sack, hamper, &c.

It is evident that in the mode of filling a corn measure, particularly in retail trade, much depends on the way in

which the corn, &c. is placed in a measure. If corn is lifted
lightly into a bushel measure, and then when the measure
is full the level of the corn is struck gently with a flat
strike and not shaken, we have perhaps the least possible
quantity in the measure, but if the corn falls into
the bushel from a height of 2 or 3 feet and is well,
shaken down, then pressed or struck with a round strike,
we have perhaps 13 per cent. more corn.   To avoid dispute
as to filling, a "hopper" is sometimes used, as in
London, Edinburgh, &c.   The form of the Board of Trade
hopper is shown as follows :—

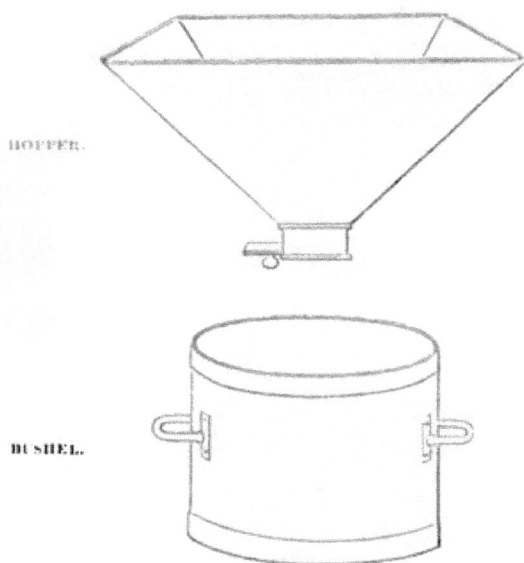

Fig. 30.—BOARD OF TRADE CORN HOPPER.

The measure to be filled is placed at a uniform distance
of 6 inches from the bottom of the hopper.

The approximate internal dimensions, in inches, of ordinary dry measures of capacity are as follows :—

	Drum Shape.		Standard Shape.	
	Depth.	Diameter.	Depth.	Diameter.
	Inches.	*Inches.*	*Inches.*	*Inches.*
Bushel - - - -	$12\frac{9}{16}$	15	$8\frac{1}{4}$	$18\frac{1}{4}$
Half-bushel - -	$10\frac{11}{16}$	$11\frac{1}{5}$	$6\frac{1}{2}$	$14\frac{3}{4}$
Peck - - - -	$9\frac{3}{8}$	$8\frac{11}{16}$	$5\frac{1}{4}$	$11\frac{11}{16}$
Gallon - - -				
Half-gallon - -				
Quart - - -	Proportionate dimensions.			
Pint - - - -				
Half-pint - -				

In testing dry measures of capacity, when grain or seed is used for the purpose of filling the measures, all measures submitted for verification should be of cylindrical form ; the internal diameter of the measure being about equal to its internal depth (Figure 31) or the internal diameter

Fig. 31.—DRUM SHAPE.　　Fig. 32.—STANDARD SHAPE.

may be about double the internal depth of the measure. (Figure 32).

In estimating the quality of corn a little instrument called a "chondrometer" is sometimes used. A given measure, as a pint, or a litre of corn, &c. is weighed by means of a portable weighing instrument. In Germany this instrument is more used than it is perhaps elsewhere and is known as a Getreideprober ; an improved form of such an instrument was introduced by the Kaiserlichen Normal Aichungs Kommission of Berlin in 1891.

The Act 39 & 40 Geo. 3. c. **81. s. 3.** (1799–1800) provided Sale of Hops. that no owner, planter, or grower of hops, shall bag any hops in any bag the weight of which bag shall be greater in proportion to the gross weight of such bag and the hops contained therein than 10 lbs., for **every 112 lbs.** of the gross weight of the **bag and the hops.**

The Act 29 & 30 **Vict. c. 37 (1866), ss. 1-22, England** and Scotland, **to prevent frauds and abuses in the trade of** hops, provides **that growers, &c. are (s.** 2) to mark **with** durable **ink or paint** on the outside of **the** bag in plain and legible figures the name of the grower, &c., and the **true** gross weight in hundredweights, quarters, and **pounds, of** each and every such bag or pocket.

The Act **36** Geo. 3. c. **85. ss. 1-3 (1795–6)** provides Miller's that every **miller** or other **person keeping a mill for the** weights. grinding of **corn shall have in such** mill a true and equal balance, with proper weights, **according** to the standard of the Exchequer, and such balance **and weights** were to be examined by the Examiners appointed **under 35** Geo. 3. c. 102.   The Chronological Table and Index **to** the Statutes (13th edition, 1896), makes the Act **to** apply to England, but is doubtful whether it applies to Scotland.

The Licensing Act (1872) 35 **& 36 Vict. c.** 94. ss. 8 and Sale of 77 (England and Ireland), provides that every person shall Intoxicating sell all intoxicating liquor which is sold by retail, and not liquors. in cask or bottle, and is not sold in **a** quantity less than half a pint, in measures marked **according** to the Imperial Standards.   In Scotland every **person** licensed to sell excisable liquors is required **9** Geo. 4. c. 58. s. 15 (1828) to sell **or** otherwise dispose of all such liquors by retail therein (except in quantities less than half-a-pint) by the gallon, quart, pint, or half-pint measure, sized according to the standard. In Ireland, also, every person licensed to sell wine by retail is to sell by the gallon, quart, pint, or half-pint measure, sized or marked according to the standard, except wine in bottle and quantities less than half a pint; (Refreshment House (Ireland) Act, 1860, 23 & 24 Vict. c. 107. s. 28).

Brewers are required (43 & 44 Vict. c. 20. ss. 28 and 46, 1880) to provide scales and weights for Excise purposes; and the Act 52 & 53 Vict. c. 42. s. 29 (1889), imposes a penalty on an excise trader keeping unjust scales and weights.

With respect to the gauging of wine or oil or other gaugeable liquors, the privileges of the City of London, or of the Lord Mayor, are reserved by s. 68. of the Weights and Measures Act, 1878.

Corn Returns.

By the Corn Returns Act, 1882, England (45 & 46 Vict. c. 37), it is provided that by the summary of quantities and prices each sort of British corn is to be computed with reference to the Imperial bushel. The inspector of corn returns is to convert into such Imperial bushel all returns made to him in any other measure, or by weight, or by a weighed measure, and in the case of weight, or weighed measure, he is to convert the same at the rate of 60 imperial pounds for every bushel of wheat, 50 imperial pounds for every bushel of barley, and 39 imperial pounds for every bushel of oats.

Sale of hay and straw. Metropolitan Acts.

The practice as to the sale of hay and straw in the metropolis is governed by section 2 of the Act 36 Geo. 3. c. 88, which provides that no hay or straw be sold within the cities of London or Westminster, or within 30 miles thereof, except in trusses. Every truss of new hay sold between 1st June and last day of August in any year, being new hay of the summer's growth of that year, is to contain the full weight of 60 lbs. and every truss of old hay of any former year's growth is to weigh 56 lbs. Every truss or bundle of straw is to contain 36 lbs., and every load of hay or straw is to contain 36 bundles or trusses. So far as regards any market through which there does not exist by law any public right of way for carts and carriages, the Act 36 Geo. 3. c. 88 was in 1834 partly repealed by the Act 4 & 5 Wm. 4. c. 21. These Acts, as well as the Act 19 & 20 Vict. c. 114, 1856, also make provision with reference to preventing the adulteration of hay and straw, and with reference to fraudulently increasing the weight of any truss of hay or straw.

In Ireland the weighing of hay and straw is, to some extent, regulated by the Weights and Measures Act, 1878, sections 76 and 77 (*see* page 31). In Scotland the Police Burgh Act, 1892, section 426, also authorises Police Commissioners to make regulations for ascertaining the weight or quantity of grain, hay, or straw, or other commodities bought within the burgh.

The law with respect to the weighing of live cattle in markets, fairs, and auction marts rested, until 1891, on the Act 50 & 51 Vict. c. 27, 1887, but in 1891 the Markets and Fairs (Weighing of Cattle) Act, 54 & 55 Vict. c. 70, was passed, which amended the principal Act of 1887. The market authority are required to provide buildings or places for weighing cattle brought for sale within a market or fair, and to keep weighing machines and weights tested by the local inspector of weights and measures. *Weighing of cattle.*

The Board of Agriculture have power under the Act to exempt certain small markets, fairs, and marts from the requirements of the Acts. The Acts also make provision with reference to statistics as to the weight and sale of cattle; and as to returns to be made to the Board of Agriculture by market authorities and auctioneers.

K

During the past century many **Acts** of Parliament have
been passed with reference to the **sale** of coal, but until
the passing of the Weights and Measures Act, 1835, coal
was largely sold by measure, and a standard measure for
coal was made in 1730, bearing the inscription "Coal
Bushel," Anno Regni Georgii Secundi **Regis** Quarto.

At present the sale of coal in England and Wales and
in Ireland is mainly regulated by Part II. of the Weights
and Measures Act, 1889.

Section 20 provides that :—

(1.) All coal shall be sold by weight only, except where
by the written consent of the purchaser, it is sold by boat
load or by waggons or tubs delivered from the colliery
into the works of the purchasers.

(2.) If any person sells coal otherwise than is required
by this section he shall be liable to a fine not exceeding 5l.
for every such sale.

The Act of 1889 provides for the detection and
prevention of fraud in the retail sale of coal by weight, and
was the first public general statute dealing with the sale
of coal in small quantities, although in some large cities
and towns useful local legislation had arisen on the subject.
In some of the poorest districts of our cities it is found to
be necessary to protect those who can only afford to buy
coals in very small quantities, as 14 lbs., and this is done by
the Act particularly by means of bye-laws made by the local
authorities of counties and boroughs, under section 28.

In London the sale of **coal** is also regulated by **local**
Acts 1 and 2 Wm. 4. c. 76 (1831) and 1 and 2 Vict. c. 101
(1837-8), which apply to places within 25 miles from the
General Post Office, including the port of London, and
Westminster. The **Act** of 1831 requires all coal, cinders,
culm, and cannel, to be sold by weight and not by measure ;

and the Acts contain provisions as to the sale of coal exceeding 560 lbs. or in any quantity under that.

The Act of 1889, except the provision requiring coal to be sold by weight only (section 20), does not extend to Scotland, or to the Scotch counties; but the Burgh Police Act, 1892 (*see* annotated edition by J. Campbell Irons, Edinburgh, 1893), contains provisions as to the prevention of fraud in the sale of coal (sections 419 to 425), similar in effect to provisions of the Act of 1889. In some burghs there is special legislation, also, dealing with the sale of small quantities of coal, as in the Glasgow Corporation and Police Act, 1895.

Coke is sold both by measure and by weight; that obtained from coal in the process of gas manufacture is largely sold retail by the chaldron measure of 36 imperial bushels; filled in all parts as nearly to the level of the brim as the size and shape of the article will admit, as "heaped measure" is now illegal.

The London local Act 14 & 15 Vict. c. 146 provides (sections 1 to 50) that, unless repugnant to the context, the word "coal," in 1 & 2 Wm. 4. c. 76. and in 1 & 2 Vict. c. 101, includes coal, coke, cinders, and culm.

Culm is a fuel which has derived its name from the geological term applied to carboniferous slatey beds in Europe.

In collieries "check-weighers" are appointed to check the weight of coal or mineral got in by the persons employed in the mines, such check-weighers being appointed by the majority of persons so employed (Coal Mines Regulation Act, 50 and 51 Vict. c. 58, 1887, section 15); and the Weights and Measures Acts apply to all the weights, measures, and gauges in use, and they are required to be tested by the local inspector once at least in every six months. *Miners' weights and measures.*

K 2

One of the most ancient local measures still in use is the Miners' and Brenners' Dish. Under the Derbyshire Mining Customs and Mineral Courts Act, 1851, the dish or measure for ore is to be provided by the "Barmaster," and it is to contain 15 pints. A Brazen Dish was at one time deposited in the Moot Hall at Wirksworth, but it is now deposited at the Museum of Practical Geology, London, the Curator of which, Mr. F. W. Rudler, has permitted the following sketch, Figure 33, to be made of this dish.

The dish contains about 14·047 Imperial pints. It is of rectangular form, and bears an inscription setting forth that "This Dishe was made the iiii day of Octob^r the iiii " yere of the Reign of Kyng henry the viii.," and that " it is to " Remayne in the Moot Hall at Wyrksworth " hangyng by a cheyne so as the Merchantes or mynours " may have resort to y same at all tymes to make the trŭ " measure after the same."

Factories and Workshops weights, &c. used in checking wages. The Factories and Workshops Act, 1878, 41 and 42 Vict. c. 16. s. 80 provides for the examination by the local inspector of weights and measures of the weights, scales, balances, steelyards, and weighing-machines used in a factory or workshop in checking or ascertaining the wages of any person employed therein in like manner as if they were used in the sale of goods.

Measurement of textile fabrics. In 1708 the Act 7 Anne, c. 13 was passed for better ascertaining the lengths and breadths of woollen cloths made in the county of York ; and in 1719 the Act 6 Geo. 1. c. 13 was also passed regulating the dimensions, not only of Scotch plaids and serges, but of stockings. Both Acts are now repealed, but the measurement, &c. of yarns is in England affected by the Act 17 Geo. 3. c. 11. (1776–7). Under the Bradford Corporation Act, 1887, the corporation were at liberty to establish, maintain, and work any

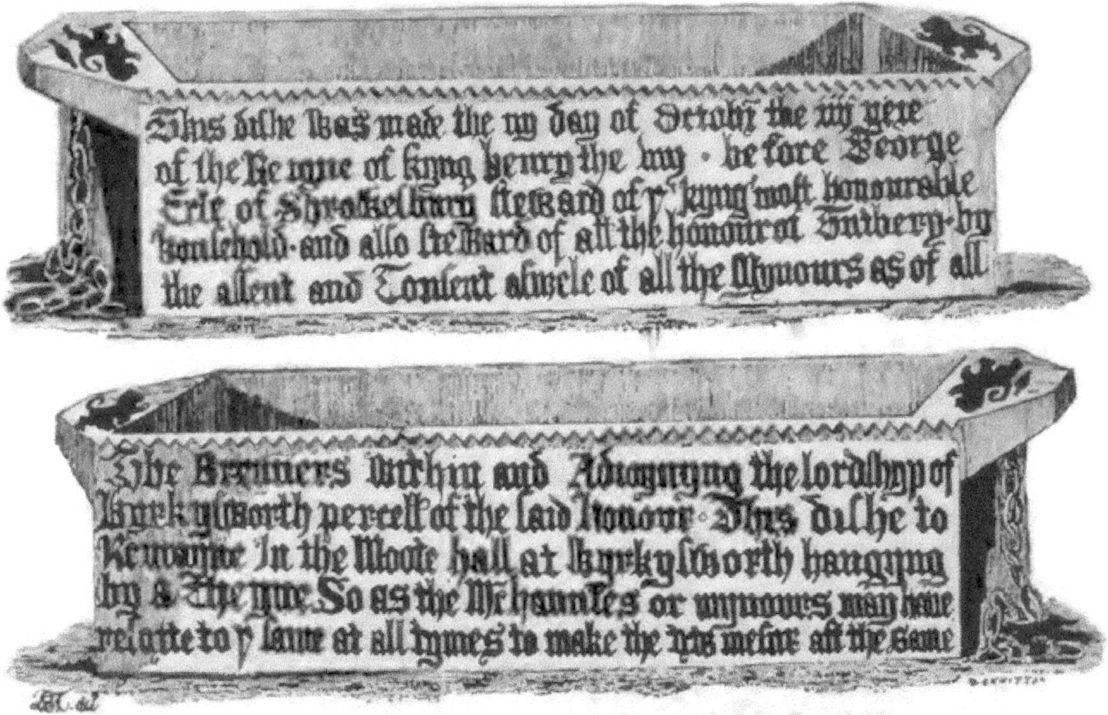

THE MINERS' STANDARD-DISH,

FOR THE WAPENTAKE OF WIRKSWORTH.

Fig. 33.

"conditioning house" or houses within the borough for the purpose of ascertaining and certifying the true weight, length, and condition of articles of trade and commerce commonly known as "tops noils and yarns" and other matters and things of similar character, and also the true weight, quality, and condition of wools, or for any other purpose of the like nature. And buyers or sellers, desirous of having any such articles tested and certified may effect such object on payment of reasonable charges by means of the Conditioning House now established at Bradford.

In textile factories an instrument known as an "Indicator" is used for the purpose of measuring and registering the work done by the mule or loom. In the cotton district, for instance, each mule-carriage is fitted with an indicator worked by a worm and toothed wheel, which automatically indicates the number of "Draws" made, or hanks of yarn spun, or the amount of work passed through the machine, and the worker, or minder of the machine, is sometimes paid according to the indications of the instrument.

The use of the indicator is recognised by the Factories and Workshops Acts (1878 and 1895); and it appears to be required that where the instrument is used in the payment of wages then it should have marked on its case the number of teeth in each wheel and the diameter of the driving roller, so that the user or worker may check the accuracy of the instrument without opening it and taking the counting apparatus to pieces. After the yarn is spun it is then weighed, and the weight of a given length ascertained by automatic indicating weighing-instruments.

For the purpose of ascertaining if the correct number of threads have been put into the calico, merchants and calico manufacturers use a "counting-glass," a small flat brass instrument containing one or two rectangular slots, $\frac{1}{2}$ and

$\frac{1}{4}$ inch respectively, and a small lens. The instrument is placed on the cloth, and the number of picks or threads within a space of $\frac{1}{2}$ or $\frac{1}{4}$ inch are counted under the lens or magnifying glass.

Some spinners, who sell cotton by weight and "counts," use also in the West Riding of Yorkshire a "wrapping-machine," which makes 80 revolutions of $1\frac{1}{2}$ yards in circumference, or 120 yards. This 120 yards of cotton is then weighed by pennyweights (dwt.) and grains. Woollen spinners also use avoirdupois dram weights for the purpose of weighing a single thread of yarn, which is then measured in yards, the different number of yards denoting also the "counts" by which the workers' wages are sometimes paid.

A yarn table, for the assistance of cotton spinners, prepared by J. Stopford and N. Gerrard (1895) shows in avoirdupois ounces and troy pennyweights and grains the length of one avoirdupois pound of cotton yarn by the weight of any number of "leas" and also the lengths from 1 to 100 hanks in the pound.

In Germany the counting of yarn is also effected by determining the number of units of length contained in a certain unit of weight. Such units vary in different countries, and at an International Congress for Uniformity in the Counting of Yarn, held in Brussels in 1874, it was proposed that an international enumeration based on the Metric System should be adopted, in which the "yarn number" for all yarns, excepting raw silk and silk thread, should be the variable weight in grammes of an invariable length in metres. The numbering of worsted yarns appears to follow the following scale :—

2 yards	=	1 thread.
40 threads	=	1 lea or warp.
7 leas	=	1 hank.

There appears to be no uniform method of folding cotton cloth at the mills, nor any generally recognised standard length of a " piece of cloth."   Usually the fold is 36 inches in length, but the Chief Inspector of Weights and Measures at Manchester states that occasionally grey cloths are folded in lengths of 36½ inches ; but rarely according to old custom in lengths of 37 inches to the yard, and that other folds are recognised in the foreign trade, as for example, metre length folds.   The mode of making up bundle yarn by dividing it into " knots " also varies ; but the growing practice appears to be to reel and sell extra hard yarns at 840 yards to the hank.

Gauge Measurements.

Cubic capacities and linear dimensions are often measured by "gauges," and some of the more important gauges now in use are as follows :— *Wire gauges.*

The gauges commonly used for measuring wires and metals are in the form of small iron or steel plates, either rectangular or circular in shape, on the edges of which notches are cut of the several required sizes, each notch bearing a descriptive number; from No. 7/0 to No. 50 as in the case of the standard gauge which was legalised by Order in Council of 23rd August 1883, or from No. 1 to No. 32, as in the case of the local gauge for sheet or hoop iron known as the South Staffordshire gauge.

The size of the gauge is measured by the distance between the parallel jaws of the notch, and, of course, it is desirable therefore that these parallel surfaces should be plane, and well finished.

STANDARD GAUGE, 1883.

Descriptive Number, B. W. G.	Equivalent in parts of an inch.	Descriptive Number, B. W. G.	Equivalent in parts of an inch.
No. 7/0	0·500	No. 23	0· 024
6/0	464	24	22
5/0	432	25	20
4/0	400	26	18
3/0	372	27	0·0164
2/0	348	28	148
0	324	29	136
1	300	30	124
2	276	31	116
3	252	32	108
4	232	33	100
5	212	34	0·0092
6	192	35	84
7	176	36	76
8	160	37	68
9	144	38	60
10	128	39	52
11	116	40	48
12	104	41	44
13	0·092	42	40
14	80	43	36
15	72	44	32
16	64	45	28
17	56	46	24
18	48	47	20
19	40	48	16
20	36	49	12
21	32	50	0·0010
22	0·028		

The above sizes, as far as No. 40, were adopted by the National Telephone Exchange Association of New York on 8th September 1885, and a table of dimensions, weight, and "resistances" of copper and iron wire, based on the above sizes, was then issued by the secretary of the

WEIGHTS AND MEASURES USED FOR SPECIAL PURPOSES.

association. In a report (1879) of a committee of the Society of Telegraph Engineers, certain sizes based on the mil. or one-thousandth of an inch were adopted.

The use of a decimal gauge is growing; in the telegraph engineering establishments of the Post Office, for instance sizes are expressed directly in thousandths of an inch, and not by descriptive numbers as in the B. W. G.

There appears to be no standard of the South Staffordshire gauge.

Besides the flat-metal gauge above referred to, other convenient forms of gauges are used, as the micrometer gauge, the French calliper gauge, American split gauge, &c. Mostly in such gauges the "inch" is followed, but milli-metre gauges are also coming into use. There are also in use local and customary gauges, as the Lancashire wire-gauge, the Yorkshire wire-gauge, Warrington wire-gauge; and the sizes of particular manufacturers have become recognised as Johnson's gauge, Stubs' standards, Holtzapffel's gauge, Wynn's gauge, &c.; of which detailed descriptions have been given by Mr. T. Hughes, C.E. (Spon. London, 1879).

Sir Joseph Whitworth has pointed out ("Papers on Engineers' gauges. Mechanical Subjects," 1882,) that formerly it did not follow of necessity that the gauges employed in any engineering shop were actually of their stated or assumed magnitudes, but that now the difficulty has been removed on the acceptance by the Board of Trade (Order in Council, 26th August, 1881) of a set of difference standard gauges, copies of which may be verified, for the use of the local authorities and engineering firms. These gauges are of cylindrical form and are made of steel. Every gauge comprises the two mechanical parts, the external gauge, which fits into the internal gauge, and they vary in size from 6 inches in diameter to 0·1 inch. Besides the cylindrical gauges, small steel flat external plane gauges are also used.

In the use of such standard gauges the Whitworth measuring machine is used, by means of which differences of one ten-thousandth part of an inch may be determined. The form of this machine is well known to engineers, and has been described by Professors Goodeve and Shelley in their book on the Whitworth Measuring Machine (Longmans, 1877).

In the manufacture of gun cartridges and revolver cartridges accuracy of measurement is important; and in cartridge and gun factories standard plugs or gauges are used whose dimensions are known to ± 0.0001 inch. At ordnance factories other accurate gauges and machines of special forms are also used in measuring external and internal diameters, as the diameters of guns, the rings shrunk on the guns, machines for measuring the eccentricity of the chamber and bore of the gun, for measuring the breech, &c.

In mechanical testings, such as testing the elasticity and strengths of metals and materials, various other forms of measuring or gauging instruments are used, as described in Unwin's " Testing of Materials of Construction," (London, 1888), results being commonly expressed in Imperial units, as the inch, the pound, the ton (Kirkealdy, "System of Mechanical Testing," London, 1891).

Screw
gauges.
The standard cylindrical gauges already referred to form also the basis of a system of screw threads largely adopted in this country and abroad for engineering purposes known as the Whitworth thread, the number of screw threads per inch being adjusted to the diameter of the screw.

As a consequence of the rapid development of steam machinery, a demand arose in mechanical work for higher accuracy, and for interchangeability in the parts of machines. This in the "screw" took form first in the production of a standard guide screw, and subsequently in the introduction of a uniform system of screw threads for engineering purposes. The development of a standard guide screw may be traced in England from the year 1828, when Mr. Bryan Donkin used such a screw with his Dividing Engine; afterwards Whitworth, and Clements of Lambeth, (who were working on Babbage's Difference Engine) produced standard guide screws of great accuracy.

WHITWORTH'S SCREW THREADS, 1841.

Diameter of Screw.	Fractional Size.	No. of Threads per Inch.	Diameter of Screw.	Fractional Size.	No. of Threads per Inch.
—	$\frac{1}{32}$	150	·750	$\frac{3}{4}$	10
—	$\frac{1}{24}$	80	·875	$\frac{7}{8}$	9
—	$\frac{1}{20}$	60	1·000	1	8
·110	--	—	1·125	$1\frac{1}{8}$	7
·120	—	—	1·250	$1\frac{1}{4}$	7
·125	$\frac{1}{8}$	40	1·375	$1\frac{3}{8}$	6
·135	—	—	1·500	$1\frac{1}{2}$	6
·150	—	—	1·625	$1\frac{5}{8}$	5
—	$\frac{5}{32}$	32	1·750	$1\frac{3}{4}$	5
·165	—	—	1·875	$1\frac{7}{8}$	$4\frac{1}{2}$
·180	—	26	2·000	2	$4\frac{1}{2}$
—	$\frac{3}{16}$	24	2·125	$2\frac{1}{8}$	$4\frac{1}{2}$
·200	—	—	2·250	$2\frac{1}{4}$	4
—	$\frac{7}{32}$	21	2·375	$2\frac{3}{8}$	4
·220	—	—	2·500	$2\frac{1}{2}$	4
·240	—	—	2·625	$2\frac{5}{8}$	4
·250	$\frac{1}{4}$	20	2·750	$2\frac{3}{4}$	$3\frac{1}{2}$
260	—	—	2·875	$2\frac{7}{8}$	$3\frac{1}{2}$
·280	—	—	3·000	3	$3\frac{1}{2}$
·300	—	—	3·250	$3\frac{1}{4}$	$3\frac{1}{2}$
—	$\frac{5}{16}$	18	3·500	$3\frac{1}{2}$	$3\frac{1}{4}$
·325	—	—	3·750	$3\frac{3}{4}$	3
·350	—	—	4·000	4	3
·375	$\frac{3}{8}$	16	4·250	$4\frac{1}{4}$	$2\frac{7}{8}$
·400	—	—	4·500	$4\frac{1}{2}$	$2\frac{5}{8}$
·425	—	—	4·750	$4\frac{3}{4}$	$2\frac{1}{2}$
—	$\frac{7}{16}$	14	5·000	5	$2\frac{1}{2}$
·450	—	—	5·250	$5\frac{1}{4}$	$2\frac{3}{8}$
·475	—	—	5·500	$5\frac{1}{2}$	$2\frac{3}{8}$
·500	$\frac{1}{2}$	12	5·750	$5\frac{3}{4}$	$2\frac{1}{4}$
·625	—	11	6·000	6	$2\frac{1}{4}$

In the United States the Whitworth system is also followed to some extent, but the Franklin Institute thread for screws and bolts, advocated by Mr. W. Sellers in 1864 and which also, like the Whitworth system, is based on the inch unit, is much used:—

SELLER'S SCREW THREAD, 1864.

Diameter of Bolt	-	¼	5-16	⅜	7-16	½	9-16	⅝	¾	⅞	1	1⅛	1¼	1⅜	1½	1⅝	1¾	1⅞
No. threads per in.	-	20	18	16	14	13	12	11	10	9	8	7	7	6	6	5½	5	5
Diameter of Bolt	-	2	2¼	2½	2¾	3	3¼	3½	3¾	4	4¼	4½	4¾	5	5¼	5½	5¾	6
No. threads per in.	-	4½	4½	4	4	3½	3½	3¼	3	3	2⅞	2¾	2⅝	2½	2½	2⅜	2⅜	2¼

In 1881 the British Association for the Advancement of Science appointed a committee for the purpose of determining a gauge for the manufacture of the various small screws used in clockwork and electrical apparatus. The Committee remark in their Report (1884) that the Swiss millimetre gauge, recommended by the Society of Art of Geneva in 1878 was very complete, but in a second report, made in 1884, it was recommended by Mr. Preece, Mr. E. Rigg, and Mr. Stroh, that the "B.A." small screw gauge should be adopted as follows. Watchmakers also follow the Swiss gauge, as explained by Professor M. Thury ("Systématique des vis horlogères," 1878).

B.A. SMALL SCREW GAUGE, 1884.

Number.	Nominal Dimensions in Thousandths of an Inch.			Absolute Dimensions in Millimetres.	
	Diameter.	Pitch.	Threads per Inch.	Diameter.	Pitch.
I.	II.	III.	IV.	V.	VI.
25	10	2·8	353	0·25	0·072
24	11	3·1	317	0·29	0.080
23	13	3·5	285	0·33	0·089
22	15	3·9	259	0·37	0·098
21	17	4·3	231	0·42	0·11
20	19	4·7	212	0·48	0·12
19	21	5·5	181	0·54	0·14
18	24	5·9	169	0·62	0·15
17	27	6 7	149	0·70	0·17
16	31	7·5	134	0·79	0·19
15	35	8·3	121	0·90	0·21
14	39	9·1	110	1·0	0·23
13	44	9·8	101	1·2	0·25

Number.	Nominal Dimensions in Thousandths of an Inch.			Absolute Dimensions in Millimetres.	
	Diameter.	Pitch.	Threads per Inch.	Diameter.	Pitch.
I.	II.	III.	IV.	V.	VI.
12	51	11·0	90·7	1·3	0·28
11	59	12·2	81·9	1·5	0·31
10	67	13·8	72·6	1·7	0·35
9	75	15·4	65·1	1·9	0·39
8	86	16·9	59·1	2·2	0·43
7	98	18·9	52·9	2·5	0·48
6	110	20·9	47·9	2·8	0·53
5	126	23·2	43·0	3·2	0·59
4	142	26·0	38·5	3·6	0·66
3	161	28·7	34·8	4·1	0·73
2	185	31·9	31·4	4·7	0·81
1	209	35·4	28·2	5·3	0·90
0	236	39·4	25·4	6·0	1·00

In the year 1888 some suggestions were given in the Gas meter Board of Trade Annual Report, page 38, with reference to gauges. the sizes of screw threads (Whitworth thread) and connecting pipes for gas meters, as shown in the following table; which suggestions, it is stated, met generally with the approval of those practically interested.

TABLE OF SIZES for the CONNECTING-PIPES and FITTINGS OF GAS METERS.

Size of Meter.	Size of Connecting-Pipe or Union of Gas-Meter.						Lining.
	Boss.			Cap.			
	Mean Diameter of External Screw.	Number of Threads per Inch.	Internal Diameter.	Mean Diameter of Internal Screw.	Number of Threads per Inch.	Height of Cap.	External Diameter, to enter Boss.
Lights.	Inch.		Inch.	Inch.		Inch.	Inch.
150	3·68	9	3·45	3·65	9	1·20	3·03
100 }80 }	3·00	11	2·30	2·95	11	1·00	2·28
60	2·45	11	2·00	2·40	11	0·80	1·98
50	2·25	11	1·80	2·24	11	0·70	1·75
30	2·05	11	1·55	2·03	11	0·70	1·53
20	1·80	11	1·42	1·75	11	0·60	1·40
10	1·45	11	1·05	1·40	11	0·60	1·03
5	1·15	14	0·83	1·10	14	0·50	0·81
3	0·98	19	0·67	0·94	19	0·50	0·65
2 }1 }	0·88	19	0·57	0·84	19	0·40	0·55
0	0·70	19	0·50	0·66	19	0·40	0·50

Gauging of needles. It is the custom of manufacturers, as well as of wholesale and retail purchasers, to gauge or buy needles by descriptive numbers. Needles are largely made in the district of Redditch, Worcester, and it would appear that the principal manufacturers recognise 16 different sizes ; the diameters of the sizes, as measured half-way between the point and the eye of the needle, being taken as follows :—

Descriptive Number.	Diameter.
1	0·044 inch.
2	0·040 ,,
3	0·036 ,,
4	0·032 ,,
5	0·028 ,,
6	0·024 ,,
7	0·022 ,,
8	0·020 ,,
9	0·018 ,,
10	0·016 ,,
11 to 16	No fixed sizes.

Gauges used in measuring ships. In the measurement of ships gauging-rods are also used, known as stanchions, or pillar and slide stanchions, which are decimally divided ; and an apparatus is used under the Board of Trade instructions known as Moorsom's Measuring Apparatus (see also Jordan's "Tabulated Weights of Angle, Tee, and Bulb Iron and Steel," London, 1896, the "Naval Architects' and Shipbuilders' Pocket Book," by C. Mackrow (16th edition, London, 1896), and other similar text books).

Official information relating to the gauges and measures used in the measurement of ships and tonnage is given in the "Instructions and Regulations relating to the Measure- " ment of Ships and Tonnage under the Merchant Shipping " Act, 1894," which were prepared and issued by the Board of Trade (Eyre and Spottiswoode, London, 1895). The

Marine Department of the Board of Trade also issues
different books of instructions relating to the **survey of
vessels**; one is entitled "Instructions as to the Survey of
" Passenger Accommodation, Master and Crew Spaces,
" Lights, and Fog Signals"; another is entitled " Regula-
" tions and Suggestions **as to** the Survey of Hull
" Equipments and Machinery of Steam Ships carrying
" Passengers." The Merchant Shipping Act, 1894, prescribes that all measurements shall **be taken** in " feet and
decimals of feet." For the purpose of international
tonnage the gross tonnage, gross deductions and net
tonnage are reduced to cubic metres.

The weight of the goods or coals contained in canal
boats is sometimes ascertained by what is known as "gauge-
weight," **that is** by gauging the depth to which **the boat**
sinks in the water when empty and when loaded. **The**
actual capacity of a boat is gauged on certain canals by the
owners of the canals for their own purposes in levying
tolls, &c. The process of gauging is simple; the boat is
brought into a testing dock and its height out of water
measured at three places on **either side of the boat by**
means of a " draught stick " **(a linear** wooden gauge or rod
graduated into 60 inches). The boat is then loaded to the
fullest extent with verified **weights** (tons avoirdupois) and
the depth of **the boat** in the water is again measured at
three different places **on** either side of **the boat.**

Gauges are used also for the purpose of testing the Fishery
measurement of nets used by fishermen under the Sea gauges.
Fisheries Acts; the form of gauge is prescribed by the
local fisheries committees. The byelaws of the Lancashire
Sea Fisheries District, for instance, provide that no person
shall use any net for taking shrimps or prawns having a
mesh through which a square gauge of three-eighths of an
inch, measured across **each** side of the square, or $1\frac{1}{2}$ inches
measured round the four sides, will not pass without
pressure when the net is wet. For sparling, mackerel, or

herring, the gauge is to be a square gauge of 1 inch measured across each side of the square, or 4 inches measured round the four sides, no person, except as above provided, is to use in fishing for sea fish any net having a mesh through which a square gauge of 1¾ inch will not pass without pressure.

The Corporation of Colchester include in their Borough plate a silver oyster used as a standard of measurement for Colne oysters.

Cask gauging.

In gauging casks the following rods or rules are used :—

A graduated head rod, for ascertaining the head diameter of the cask. This gauge serves also as a sliding rule for calculating the quantity of liquid in the cask.

A bung rod and dipping rod, for finding the bung diameter and diagonal of the cask.

Long callipers, used for finding the internal length of a cask from head to head. Short or cross callipers, for finding the external diameter of a cask. Stave gauge, for finding the thickness of the stave of a cask.

Measurements are generally made in feet and inches, rules for gauging casks are, as is well known, laid down in works on mensuration, &c. ; for revenue purposes, excise officers are largely guided by rules given in the "Excise Officers' Manual."

The contents of beer casks are sometimes determined by weight.

The late Sir J. Lubbock, F.R.S., in a paper on "Cask-Gauging," which was reprinted in 1834 by Charles Knight from the "London and Edinburgh Philosophical Magazine," states that Kepler was the first to endeavour to reduce the art of "gauging" to accurate principles in his work entitled "Nova Stereometria Doliorum Vinariorum. 1615."

Gauging timber.

In gauging timber quantities are usually measured and expressed in cubic feet run.

The gauging or measurement of timber at our ports is undertaken by timber measurers appointed by the

Directors of the Customs Annuity and Benevolent Fund, who are known as Customs Fund measurers. The rules for gauging are the same as those followed by the Customs Department before the duties on wood goods were abolished in 1866. The measurements by the Customs Fund officers appear in 1892 to have amounted to 600,000 loads for freight and sale purposes. The marking or scribing of the cubic contents on bulk timber does not appear to be compulsory by Act of Parliament, and it is the practice of the measurers to scribe all large timber with a number corresponding to the records in their books of measurement.

As an instance of the classification of timber for canal charges or tolls, it is sometimes provided in Canal Acts that "Forty cubic feet of oak, mahogany, teak, beech, "greenheart, ash, hickory, ironwood, baywood, or other "heavy timber, and 50 cubic feet of poplar, larch, fir, elm, "birch, lancewood, walnut, or other light timber, other "than deals, battens, or boards, and 66 cubic feet of "deals, battens, and boards, shall be charged for as one "ton." A load of timber is generally taken as equal to 40 cubic feet of unhewn timber, or 50 feet of squared timber, or 600 superficial feet of 1-inch planks or deals.

Gauges for measuring the height of water in a river or lake are of ancient origin, as the Nilometer of Egypt, and in modern times consist not only of a graduated rod or post but include a float and water tube, with magneto-electric means for indicating and recording the height of the water. *Water gauges or meters.*

In the sale of water meters or gauges are used, but their use is not regulated by any public general Act of Parliament. Local authorities and water companies, however, sometimes test water meters for use in their own district. Such instruments generally register in gallons, and are classified as positive meters and inferential meters, the former

L.

actually measuring the water passed, in a vessel of known
capacity, while inferential meters infer the quantity passed
by the revolutions imputed to a fan or turbine by the
water as it passes through the meter ("The Water-Meter,"
W. G. Kent, London, 1892).

Slide rules.   Forms of gauges known as "slide rules" are used
in the mechanical measurement of work, or in checking
measurements where approximate accuracy is needed, and
where it is desired to solve at sight questions which depend
on ratio. In an extended slide rule arranged by Major-
General Hannyngton the scale was so graduated in parts that
a radius of 60, 120, and 240 inches might be obtained, and
such a scale is said therefore to be equal in power to straight
gauges of 10, 20, and 40 feet respectively. Gauges or slide
rules, graduated to meet the requirements of particular
trades, are also used, as the engineer's slide rule, rules for
the calculation of embankment work, ditches, canals, and
fortifications, or for setting out railway curves. Slide
rules for logarithms, and sines, areas, diameters, and
circumferences ; slide rules of involution, giving the
powers and roots of any given number, are also made.
Rules are commonly used for the measurement of brick-
work, curbing-stone, &c., and like the ordinary carpenter's
slide rule, combine scales of parts of an inch with the
logarithmic scales. In timber measurement special rules
are used for finding the superficial or cubic contents of
round and unequal-sided timber (St. Petersburg standard, &c.). Slide rules are used also for sheet-iron and
steel measurements, or for rolling-mills, showing the lengths
of bars of any size and form ; for measuring volumes of
gas, temperature effects, and barometric heights ; for finding
the weights of materials from their specific gravities ; for
determining breaking strains, and for general use in
mechanical testings. Compound slide rules, arranged on
"Gunter's lines," are used in solving problems in trigono-
metry and navigation.

Fig. 34.—INSPECTOR'S OFFICES (ENTRANCE), SOUTH CENTRAL STATION, NEWINGTON, LONDON.

The denominations of weights and measures used for the special purposes of trade are shown in Appendix 1. The number of weights and measures stamped in the United Kingdom is very large (3,500,000 weights, and 4,750,000 measures, annually), and the amount of verification and inspection work done in our manufacturing centres and populous counties is probably larger than that done in other similar districts abroad. For instance, in Birmingham alone, 19,390 weighing instruments were verified and stamped during 1895-6, (Annual Report, Chief Inspector, 1896, p. 4), and similar returns of weights, &c., stamped, were made in other districts. Another instance of one of our large inspectoral districts is that of the London County Council, whose staff of 75 inspectors and assistants made, during the past year, 43,530 inspections of premises where weights and measures were used for trade, and examined also 4,640 costermongers stalls, barrows, &c. (Annual Report, Chief Officer, Public Control Department, Mr. A. Spencer, 1895-6, p. 9).

For the accommodation of their inspectors, local authorities provide offices and instrumental equipments, and some of the principal offices are those in cities, like Bristol, Glasgow, Liverpool, Leeds, London, Manchester, Newcastle, Sheffield, &c. In Fig. 34 there is shown an exterior view of the South Central Weights and Measures Office, provided by the London County Council; other stations being also provided by them at Marylebone, Clerkenwell, Bethnal Green, Battersea, Blackheath, and Paddington. At the South Central Station (Union Road, Newington) separate offices, with standards and testing and stamping appliances, are set apart for the verification of the various forms of weights, measures, and weighing-instruments. The entrance into the offices is shown on the right of Fig. 34; outer lobbies and a receiving room are provided for the convenience of the public who bring their weights, &c., to be verified; and on the left of the entrance gates, one side of the stabling is shown, in which are kept the horses and vehicles used in the outdoor inspection. In Fig. 35 an interior view is shown of the separate office used for testing publican's glass measures (now so largely used in place of pewter measures), the verifying officers being seated at the testing benches, at which they are comparing the glasses with standard measures verified by the Board of Trade; the stamp or mark of verification being placed on each verified glass by means of a "sand-blast" machine.

Of course such extensive offices as those above referred to are not generally required in counties and boroughs, and in many instances the inspector's accommodation is limited to one or two rooms, situated at or near the county offices or local municipal buildings.

APPENDIX I.

Legal denomination.	Abbreviation allowed.

MEASURES.

Measures of length :

100 feet - - -	100 ft.
66 feet, or a chain of 100 links	Chain.
Rod, pole, or perch - -	Pole.
10 feet - - - -	10 ft.
6 ,, - - -	6 ,,
5 ,, - - -	5 ,,
4 ,, - - -	4 ,,
3 ,, - - -	3 ,,
2 ,, - - -	2 ,,
18 inches - - -	18 in.
1 foot - - -	1 ft.
Yard - - -	1 yd.
Half-yard - - -	$\frac{1}{2}$,,
Quarter-yard - - -	$\frac{1}{4}$,,
Eighth of a yard - -	$\frac{1}{8}$,,
Sixteenth of a yard - -	$\frac{1}{16}$,,
Nail - - -	Nail.
Inch - - -	In.

Measures of capacity :

Liquid measures :

Thirty-two gallons - -	32 gallons.
Thirty-one ,, - -	31 ,,
Thirty ,, - -	30 ,,
Twenty-nine ,, - -	29 ,,
Twenty-eight ,, - -	28 ,,
Twenty-seven ,, - -	27 ,,
Twenty-six ,, - -	26 ,,
Twenty-five ,, - -	25 ,,
Twenty-four ,, - -	24 ,,
Twenty-three ,, - -	23 ,,
Twenty-two ,, - -	22 ,,
Twenty-one ,, - -	21 ,,
Twenty ,, - -	20 ,,
Nineteen ,, - -	19 ,,
Eighteen ,, - -	18 ,,
Seventeen ,, - -	17 ,,
Sixteen ,, - -	16 ,,

Legal denomination.	Abbreviation allowed.

MEASURES—*cont.*

Fifteen gallons - - -	15 gallons
Fourteen ,, - - -	14 ,,
Thirteen ,, - - -	13 ,,
Twelve ,, - - -	12 ,,
Eleven ,, - - -	11 ,,
Ten ,, - - -	10 ,,
Nine ,, - - -	9 ,,
Eight ,, - - -	8 ,,
Seven ,, - - -	7 ,,
Six ,, - - -	6 ,,
Five ,, - - -	5 ,,
Four ,, - - -	4 ,,
Three ,, - - -	3 ,,
Two ,, - - -	2 ,,
One gallon - - - -	1 gallon
Half-gallon - - -	$\frac{1}{2}$,,
Quart - - - -	Quart.
Pint - - - -	Pint.
Half-pint - - -	$\frac{1}{2}$ pint.
Gill - - - -	Gill.
Half-gill - - -	$\frac{1}{2}$ gill.
Quarter-gill - - -	$\frac{1}{4}$ gill.
Dry measures :	
Four bushels - - -	4-bushel.
Bushel - - - -	Bushel.
Half-bushel - - -	$\frac{1}{2}$ bushel.
Peck - - - -	Peck.
Gallon - - - -	Gallon.
Half-gallon - - -	$\frac{1}{2}$ gallon.
Quart - - - -	Qrt.
Pint - - - -	Pt.
Half-pint - - -	$\frac{1}{2}$ pt.
Apothecaries' measures :	
40 fluid ounces, to half a fluid ounce - - - -	40 fl. : oz., to $\frac{1}{2}$ fl. : oz.
16 fluid drachms, to half a fluid drachm - - -	16 fl. : dr., to $\frac{1}{2}$ fl. : dr. } or symbols.
60 minims, to 1 minim - -	60 min. ($\mathfrak{m}$), to 1 min. ($\mathfrak{m}$).

WEIGHTS.

Avoirdupois weights :	
Cental, or 100 pounds - -	Cental, or 100 lb.
56 pounds, or half-hundred weight - - -	56 lb., or $\frac{1}{2}$ cwt
28 pounds, or quarter-hundred-weight - - -	28 ,, or $\frac{1}{4}$,,

Legal denomination.	Abbreviation allowed.

WEIGHTS—cont.

Legal denomination.	Abbreviation allowed.
14 pounds, or stone	14 lb., or stone.
7 ,,	7 ,,
4 ,,	4 ,,
2 ,,	2 ,,
1 pound, or 7,000 grains	1 ,,
8 ounces, or half pound	8 oz., or ½ lb.
4 ,, or quarter pound	4 ,, or ¼ lb.
2 ,,	2 ,,
1 ounce, or 437½ grains	1 ,,
8 drams, or half-ounce	8 dr., or ½ oz.
4 ,, or quarter-ounce	4 ,, or ¼ ,,
2 ,,	2 ,,
1 **dram**	1 ,,
½ ,,	½ ,,
Pennyweights :	
480 grains, **or** 20 pennyweights	480 gr., or 20 dwt.
240 ,, 10 ,,	240 ,, or 10 ,,
120 ,, 5 ,,	120 ,, or 5 ,,
72 ,, 3 ,,	72 ,, or 3 ,,
48 ,, 2 ,,	48 ,, or 2 ,,
24 ,, 1 pennyweight	24 ,, or 1 ,,
Troy weight :	
Decimal troy ounce bullion weights :	
500 ounces, troy	500 oz. tr.
400 ,,	400 ,,
300 ,,	300 ,,
200 ,,	200 ,,
100 ,,	100 ,,
50 ,,	50 ,,
40 ,,	40 ,,
30 ,,	30 ,,
20 ,,	20 ,,
10 ,, or 4,800 grains	10 ,,
5 ounces, to 0·001 ounce	5, to 0·001.
Apothecaries' weight :	
10 **ounces**	10 oz. tr.
8 ,,	8 ,,
6 ,,	6 ,,
4 ,,	4 ,,
2 ,,	2 ,,
1 ounce, or 480 grains	1 ,,
{ 4 drachms	℥ iv.
{ or half an **ounce**	½ oz tr.
2 drachms	℥ ij
1 drachm	℥ i
2 scruples	℈ ij

Legal denomination.	Abbreviation allowed.

WEIGHTS—*cont.*

Legal denomination.	Abbreviation allowed.
{ 1½ scruples	3 ℈ß
or half a drachm	3 ℈ß
1 scruple	℈ i
half a scruple	℈ ß
6 grains	6
5 ,,	5
4 ,,	4
3 ,,	3
2 ,,	2
1 grain	1
half a grain	½
Grain weights :	
4,000 grains	4,000 gr.
2,000 ,,	2,000 ,,
1,000 ,,	1,000 ,,
500 ,,	500 ,,
300 ,,	300 ,,
200 ,,	200 ,,
100 ,,	100 ,,
50 grains, to 0·01 grain	50, to 0·01.

Electrical standards, *see* page 75.

APPENDIX 2.

Representative Forms of Weighing-Instruments used in Trade.

Scale Beam with Swan-neck Ends (Plate 1, Figs. 1).
Scale Beam with Box Ends (Plate 1, Figs. 2).
Scale Beam with Dutch Ends (Plate 1, Figs. 3).
Grocers' Scale Beam (Plate 1, Figs. 4).
Scale Beams with continuous knife-edge Bearings (Plate 1, Figs. 5).
Bullion Balance (Plate 1, Figs. 6).
Accelerating Counter Machine (Plate 1, Figs. 7).
Vibrating Counter Machine (Plate 1, Figs. 8).
Inverted Counter Machine (Plate 1, Figs. 9).
Stays of Counter Machines (Plate 1, Figs. 10).
Coal Steelyard (Plate 2, Fig. 1).
Roman Steelyard (Plate 2, Fig. 2).
Steelyard Counter Machine (Plate 2, Figs. 3).
Automatic Machine (Plate 2, Fig. 4).
Platform Machine (Plate 2, Figs. 5).
Coal Platform Machine (Plate 2, Figs. 6).
Skew Lever Weighbridge (Plate 2, Figs. 7).
Truck Weighbridge (Plate 2, Figs. 8).
Double Truck Weighbridge (Plate 2, Fig. 9).
Crane Machine (Plate 2, Fig. 10).
Spring Balance (Plate 2, Fig. 11).

INVERTED COUNTER MACHINE.

ELEVATION.

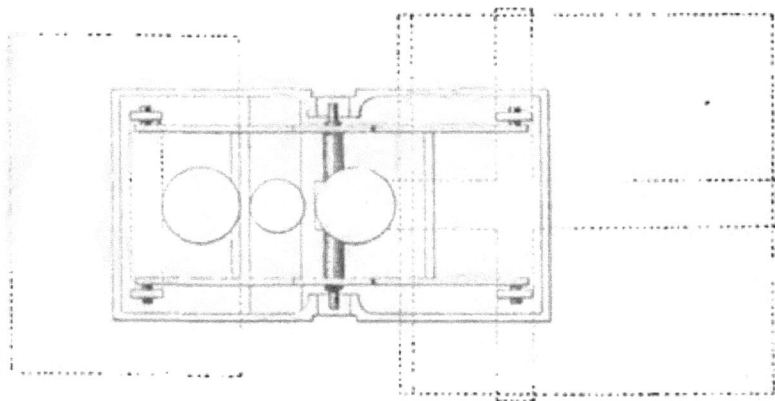

PLAN.

Scale : 7/8 th.

INDEX.

A.

Acre, 27, 68.
Acts, 44, 46, 127 to 137.
Africa, 44, 66.
Agriculture, Board of, 65.
Air, 86.
Airy, Mr. W., 60.
Airy, Sir G. B., 58.
Alcohol, 81.
Amsterdam, 36.
Ampère, 75.
Ancient standards, 10, **14, 16, 51.**
Anglo-Saxon, 17, 18.
Annoyance juries, 55.
Apothecaries, 14, 21, 115.
Arabia, 18, 21.
Arago, 103.
Arbuthnot, Dr., 16, 18.
Assay, 38, 122, 125, 126.
Assyria, 17, 21.
Australia, 44, 45.
Avoirdupois, 1, 4, **7, 13, 18.**

B.

Babylon, 20.
Baily, Mr., 4.
Balances, 61, **62, 93.**
Bankers, 123.
Bannister, Mr. R., 80.
Barley, 24, 128, 132.
Barometer, 79.
Bate, Mr., 81.
Beanmé, 82.
Benoit, Dr., 98, 104.

B.

Bernard, Dr., 24.
Bird, Mr., 3.
Birmingham, 122, 151.
Bradford, 137.
Bread, 126.
Brenner's dish, 136.
Broch, Dr., 98.
Bullion, 125.
Burmah, 40.
Burner, 85.
Bushel, Imperial, 8, 9, 88, 130, 132.
    ,,   Winchester, 11, **12,** 15.

C.

Canada, 42.
Canals, 147.
Capacity, 7, 88, 130.
Cape of Good Hope, 44, 66.
Carat, 21, 125, 126.
Cardew, Major P., 75.
Carysfort, Lord, 2.
Casks, 57, 148.
Cattle, 25, 133.
Cental, 128.
Centigrade, 78.
Chains, 68, 70.
Channel Isles, 32.
Chappuis, Dr. 98.
Chemists, 82, 115.
Chester, 122.
Chisholm, Mr., **20.**
Chondrometer, **130.**
Choppin, 29.
Clarke, Col., 73, 110.

Cloth, 57, 136.
Coal, 134.
Coinage, 24, 118, 121.
Coin Weights, 19, 118.
Coke, 135.
Collieries, 134, 135.
Comparators, 90.
Constabulary, 13, 49.
Coopers' Co., 56.
Corn, 34, 35, 127, 132.
Cotton, 138.
Cran, 30.
Cubic Measure, 84, 86, 88.
Cumberland, Bishop, 20.

D.

Decimal, 13, 43, 112, 124, 141.
Decimeter, 22, 110.
De Morgan, Professor, 22.
Deville, M., 98.
Diamond, 126.
Dividing Machine, 92.
Domesday Book, 18.
Dublin, 3, 49, 72, 122.

E.

Edgar, King, 15, 17.
Edinburgh, 3, 28, 72, 122.
Education, 112.
Egypt, 17.
Electrical Standards, 75.
Elizabeth, Queen, 2, 4.
Ell, 23, 27, 112.
Engineers, 60, 68, 141.
Equivalents, 101.
Everett, Professor, 74.
Examiners, 47.
Exchequer, 2, 5, 14, 31, 120.

F.

Factories, 136.
Fahrenheit, 78.
Fathom, 71.
Fiar prices, 26.
Firlot, 26.

Fishery gauges, 30, 147.
Fleetwood Bishop, 17.
Foot, 18, 36.
Founders' Co., 56.
Francœur, M., 86.
Fruiterers' Co., 57.

G.

Gallon, 7, 8, 11, 12, 35, 59.
Gas, 83, 145.
Gauges, 139, 147.
Gay Lussac, 83.
Germany, 17.
Gold, 7, 24, 98, 118, 121, 125.
Glasgow, 72, 151.
Goldsmiths' Co., 56, 122.
Gould, Dr., 104.
Graham, Mr., 3.
Grain, 13, 23, 125, 127.
Greaves, Professor, 17.
Greece, 16, 21.
Guillaume, Dr., 74, 98.
Gunter, Professor, 70.

H.

Hand, 23.
Hannyngton, M.-General, 150.
Harbours, 58.
Harris, Mr., 3.
Hay, 132.
Heaped measure, 128.
Henry VII., 2, 4, 10.
Hopper, 129.
Herschel, Sir J., 22.
Houses of Parliament, 3, 4, 9.
Human body, 23.
Hydrometer, 80.

I.

Imperial system, 1, 25, 30, 36, 43, 44, 128.
   „   standards, 1, 5, 7, 8, 9.
India, 36, 66.
Inspection, 46.

Inspectors, **3, 47, 62, 84, 151.**
Institution, Engineers', 60.
International Committee, 96.
Intoxicating liquors, **131.**
Ireland, 13, 25, 30, **49,** 52, 127, **134.**

J.

Jersey, 32.
Jewel Tower, 93, 122.
Jewellers, 125.
Juries, 54, 55.

K.

Kater, Capt., 2, 103.
Kelvin, Lord, 77.
Kilogram, 22, **96, 98.**
Klafter, 67.

L.

Lanark, 26, 28.
Land measure, 65, 67.
Libra, 18.
Light, 23, 85.
Linear measure, 1, 65, 90.
Linlithgow, 28.
Litra, 18.
Litre, **110.**
Local measure, 25, 128.
    „   practice, 13, 46.
London, 3, 55, 127, 132, 134, **151.**

M.

Malay, 40.
Manchester, 72, **151.**
Man, Isle, 33.
Manors, 54.
Marc, 32.
Martin, Major, 38.
Maps, 67.
Markets, 52, 53.
Mass, 88.

Matthey, **Mr. G., 98.**
Mendeléeff, Professor, 89, 94.
Merchandize marks, 55.
Merchant Taylors' Co., 57.
Meridian, 22, 110.
Meter, 75, 83, **145, 149.**
Metre, 22, 23, **34, 96, 104.**
Metric system, **2, 36, 43, 96, 109,**
    113.
Metrological Society, 36.
Michelson, Professor, 23.
Microscope, 90.
Mile, 17, 27, 67, 101.
Miller, Professor, 4, 19.
Mines, 70, 135.
Mural standards, 72.
Music, 22.
Moneyers, 19, **20, 120.**

N.

Nail, 23.
Natal, 44.
Natural constant, 21.
Nebuchadnezzar, 21.
Needles, 146.
New South Wales, 44.
Newton, Sir I., 67.
New Zealand, 44.
Norman, 17, 19, 32, **120.**
Norwich, 15.

O.

Ohm, 75.
Ordnance survey, 65.
Orders in Council, 59, 72, 88, 128,
    139, 141.
Origin of standards, 1, 16, **21, 109.**
Ounce, 18, 20, 124.
Ox, 25.

P.

Palgrave, Sir R., 3.
Parliament, 3, 4, 9.
Pawnbrokers, 124.
Pendulum, 21, 73.

Penny, 24, 119, 124.
Pentane, 86.
Peru, 67.
Petrie, Professor, 20.
Pharmaceutical Society, 117.
Phœnicia, 17
Photometer, 85.
Physical standards, 21.
Pint, 14, 18, 26, 106.
Playfair, Dr., 2.
Plumbers' Co., 57.
Poole, Mr. Lane, 119.
Pound. See Avoirdupois, Troy.
Pyramid, 16.
Pyx, 120, 121.

Q.

Quarter, 26, 106, 128, 131.
Queipo, Don V., 17.

R.

Refiners, 125.
Ribands, steel, 71.
Ridgeway, Professor, 24.
Rhineland foot, 36.
Rogers, Professor, 104.
Rome, 17, 18, 19, 20.
Royal Mint, 4, 119, 121, 125.
    „   Observatory, 4, 73.
    „   School of Mines, 125.
    „   Society, 3, 4, 8, 89, 103.
Ruding (Coinage), 20, 25.
Rupee, 39.

S.

Saccharometer, 81.
Sand-blast, 152.
Saxon, 17, 18, 19, 20, 121.
Schools, 112.
Scientific instruments, 74.
Scotland, 18, 19, 25, 30, 48, 127, 135.
Screws, 142.
Scully, Lieut.-Col., 38.
Sextarius, 18.
Sheepshanks, Rev. R., 4, 78, 79.

Sheffield, 122, 151.
Shekel, 20.
Ships, 72, 146.
Slide rules, 150.
Smyth, Professor P., 16.
South Kensington Museum, 73.
Spain, 66.
Special weights, &c., 115.
Stamp, 14, 50, 84, 85, 119.
Standards Commission, 5, 58.
Standards. See Imperial, &c.
Standards Office. See Trade, Board of.
Stanley, Dean, 120, 122.
Stas, Professor, 98.
Statutes, 1, 44, 46, 127 to 137.
Steel measures, 67, 69.
Sterling, 24.
Stirling, 27, 28.
Stone, 24, 128.
Stoup, 26.
Strachey, Sir R., 39.
Straits Settlements, 40.
Straw, 132.
Surveying, 65, 68.
Swinton, Lord, 19.

T.

Tapes, 69, 71.
Tasmania, 45.
Taylor, Mr. J., 16.
Temperature, 67, 77, 92.
Textile fabrics, 136.
Timber, 148.
Tittmau, Mr., 104.
Toise, 67, 104.
Tola, 38, 39.
Tower, 19, 20, 122.
Trade, Board of, 5, 13, 50, 58, 60, 75, 83, 85, 96, 105, 121, 122, 146, 153.
Trade Weights, &c., 46, 59, 112.
Trafalgar Square, 72.
Tralles (Hydrometer), 83.
Tresca, M., 98.

Trigonometrical Survey, **65.**
Tron, 19, 26, 27.
Troy, 3, 4, 13, 19, 26, 27, 35.

U.

Unciæ, 18.
United States, 34, 81.
Units, 1, 17, 21, 74, 99, 142.

V.

Verification, 46.
Victoria (Australia), 45.
Volt, 75, 77.

W.

Wakefield, 54.
Wales, 44, 48, 128, 134.
Walker, General, 104.

Warden, 12.
Watchmakers, 144.
Water, 8, 9, 26, 88, 149.
Weighing instruments, 31, 59, 93.
Weight, 88.   *See* Avoirdupois, Troy.
West Indies, 45.
Westminster, 1, 87, 120, 122, 132, 134.
Wheat, 24, 25, 127.
Whitworth, Sir J., 141.
Wilson, Mr. A., 18, 19.
Winchester, 3, 7, 11, 15, 35, 36.
Wire gauge, 139.
Wollaston, Dr., 2.
Wool, 136.
Wybard, Dr., 89.

Y.

Yard, 1, 5, 6, 10, 21, 23, 31, 90, 103.
Young, Dr. T., 2.

www.ingramcontent.com/pod-product-compliance
Lightning Source LLC
Chambersburg PA
CBHW021709210326
41599CB00013B/1588